Gene Transfer to Animal Cells

Gene Transfer to Animal Cells

R.M. Twyman

Department of Biological Sciences,
University of York, York, UK

BIOS Scientific Publishers
Taylor & Francis Group

A CIP catalogue record for this book is available from the British Library.

ISBN 1 8599 6204 1

Garland Science/BIOS Scientific Publishers
4 Park Square, Milton Park, Abingdon, Oxon OX14 4RN, UK and
270 Madison Avenue, New York, NY 10016, USA
World Wide Web home page: www.garlandscience.com

Garland Science/BIOS Scientific Publishers is a member of the Taylor & Francis Group.

Distributed in the USA by
Fulfilment Center
Taylor & Francis
0650 Toebben Drive
Independence, KY 41051, USA
Toll Free Tel.: +1 800 634 7064; E-mail: taylorandfrancis@thomsonlearning.com

Distributed in Canada by
Taylor & Francis
74 Rolark Drive
Marborough, Ontario M1R 4G2, Canada
Toll Free Tel.: +1 877 226 2237; E-mail: tal_fran@istar.ca

Distributed in the rest of the world by
Thomson Publishing Services
Sherlton House
North Way
Endover, Hampshire SP10 5BE, UK
Tel.: +44 (0)1264 332424; E-mail: salesorder.tandf@thomsonpublishingservices.co.uk

Library of Congress Cataloging-in-publication Data

Twyman, Richard M.
    Gene transfer to animal cells / R.M. Twyman.
        p. ; cm.
    Includes bibliographical references and index.
    ISBN 1-85996-204-1
    1.   Genetic engineering. 2.   Transfection.
        [DNLM: 1.   Gene Transfer Techniques. 2.   Animals, Genetically Modified.
    3.   Gene Targeting. 4.   Genetic Markers. 5.   Genetic Vectors. 6.   Transgenes.
    QZ 52 T975g 2005] I. Title.
        QH442.6.T98 2005
        660.6'5--dc22
                                            2004023190

Production Editor: Georgia Bushell
Typeset by Phoenix Photosetting, Chatham, Kent, UK
Printed and bound by TJ International Ltd, Padstow, Cornwall

Contents

Abbreviations

AAV	adeno-associated virus	IMP	inosine monophosphate
ADA	adenosine deaminase	IRES	internal ribosome entry site
AMP	adenosine 5′-monophosphate	LAT	latency-associated transcript
ATT	adenosine 5′-triphosphate	LCR	locus control region
BAC	bacterial artificial chromosome	LTR	long terminal repeat
BIV	bovine immunodeficiency virus	MLV	murine leukemia virus
BPV	bovine papillomavirus	MMTV	murine mammary tumor virus
CAT	chloramphenicol acetyltransferase	NPV	nuclear polyhedrosis virus
		PEI	polyethyleneimine
CHO	Chinese hamster ovary	PEG	polyethylene glycol
CID	chemically induced dimerization	PFU	plaque-forming unit
CMV	cytomegalovirus	RAGE	recombinase activation of gene expression
DHFR	dihydrofolate reductase		
DMSO	dimethylsulfoxide	REMI	restriction enzyme-mediated integration
dTTP	deoxythymidine triphosphate		
EBV	Epstein–Barr virus	RIGS	repeat-induced gene silencing
EGC	embryonic germ cell	RSV	Rous sarcoma virus
EIAV	equine infectious anemia virus	SCID	severe combined immunodeficiency
ES	embryonic stem		
FIV	feline immunodeficiency virus	SEAP	secreted alkaline phosphatase
GFP	green fluorescent protein	SFV	Semliki Forest virus
GMP	good manufacturing practice	SIN	Sindbis virus
GMP	guanosine 5′-monophosphate	SIV	simian immunodeficiency virus
HIV	human immunodeficiency virus	TK	thymidine kinase
HSV	herpes simplex virus	TNF	tumor necrosis factor
ICSI	intracytoplasmic sperm injection	UTR	untranslated region
IFN	interferon	UV	ultraviolet
IMAC	immobilized metal-affinity chromatography	VSV	vesicular stomatitis virus
		YAC	yeast artificial chromosome

Preface

Gene transfer methods for animal cells were first developed in the early 1960s, when researchers working with cultured mammalian cells sought ways to enhance the efficiency of infection with viral DNA and RNA. Since then, techniques for the introduction of exogenous genetic material have diversified and improved significantly, and they now underpin much of the current research in molecular and cellular biology as well as forming the basis of many applied aspects of biotechnology. A wide variety of methods has been described for gene transfer into animal cells, including chemical and physical delivery techniques, gene transfer using viral vectors and, most recently, gene transfer using bacterial vectors. These techniques are used not only to add new DNA to animal cells, they also form the basis of methods which allow random or targeted gene disruption and other forms of genome modification. In the last few years, the delivery of RNA into animal cells has become increasingly important as researchers have turned to RNA interference as a method for the inactivation of specific genes.

Although initially developed for the transformation of cells in culture, many gene transfer techniques have also been adapted to allow the transformation of cells in vivo, opening up new medical applications such as gene therapy and the use of DNA vaccines. Furthermore, gene transfer to germ cells, gametes or early embryos facilitates germline transformation and thus allows the production of transgenic animals, in which every cell is genetically modified. More recently, transgenic animals have also been produced through the manipulation of somatic cells followed by nuclear transfer into enucleated eggs. As gene transfer methods have matured, so we have come to understand in more detail the factors that influence the delivery, behavior and expression of exogenous genes, helping to increase the efficiency of gene transfer, enabling more control over the delivery system and facilitating the precise regulation of transgene expression. Over the last 40 years, gene transfer to animal cells has thus evolved from a rather hit-and-miss affair to an independent discipline in which precise goals can be achieved with high levels of efficiency.

To the researcher faced with the myriad of different delivery methods, host–vector systems and expression strategies, it can be difficult to make informed choices as to the most suitable strategy for any particular set of experimental goals. The aim of this *Advanced Methods* book is to provide an overview of these many different techniques, to summarize their advantages and disadvantages, and to provide guidance as to what can and cannot be achieved with each method. After a brief introductory chapter, which summarizes contemporary gene transfer technologies and provides an overview of the different selectable and scorable marker genes used with animal cells, Chapters 2–4 cover the different gene transfer methods in more detail. Chapter 2 considers chemical and physical transfection methods, Chapter 3 focuses on the relatively recent development of cell-based delivery systems and Chapter 4 provides an overview of the different viral vectors that are available for gene delivery. These chapters cover all the different vector/delivery systems and their applications in terms of transient vs stable expression, integration vs episomal maintenance, host range, vector construction and major applications. Chapter 5 extends the discussion of these vector/delivery systems in the context of producing transgenic animals. Whereas the first half of the book focuses on gene transfer methods, the second half is more goal-orientated and considers the different ways in which gene transfer can be applied to achieve different aims. Chapter 6 covers the use of homologous recombination and site-specific recombination to disrupt or modify endogenous genes and genomes. Chapter 7 provides an overview of available strategies for controlling transgene expression, including the use of inducible expression systems. Chapter 8 briefly considers some of the principles of transgene behavior which can generate unexpected results in gene transfer experiments, such as the impact of position effects, dosage effects, genetic background and transgene structure. Chapter 9 focuses on the use of gene transfer to achieve the targeted or random inactivation of endogenous genes, e.g. with antisense RNA, ribozymes, RNA interference or specific recombinant antibodies. Finally, Chapter 10 briefly outlines some of the major applications of gene transfer in animal cells, including the study of gene function and regulation, the commercial synthesis of recombinant proteins, the improvement of domestic animals, disease modeling and gene therapy. Selected methods are included in some of the chapters to provide the reader with examples of how the range of methods discussed in the book could be applied, but this is not intended to be comprehensive or exhaustive coverage because the range of systems, methods and applications discussed in the main text would make this an impossible goal. For readers interested in the fine details of individual methods, there are many specialized texts available on particular expression systems or applications, and these are mentioned in the further reading section included at the end of each chapter.

This book would not have been possible without the help and support of the team at Garland-BIOS, particularly Chris Dixon and Nigel Farrar. I would like to thank all at the Fraunhofer Institute of Molecular Biology and Applied Ecology, the Technical University in Aachen and the Department of Biological Sciences, University of York. Very special thanks are due to Ricarda Finnern, RWTH Aachen, for patient editing and attention to detail.

This book is dedicated with love to my parents, Peter and Irene, to my children, Emily and Lucy, and to Paula who makes it all worthwhile.

Richard Twyman

Basic principles of gene transfer

1

1.1   Introduction

The delivery of DNA into animal cells is a fundamental and established procedure, which is used very widely both in basic research and applied biology. In research, the transfer of DNA into animal cells is an indispensable tool for gene cloning, the study of gene function and regulation, and the production of small amounts of recombinant proteins for analysis and verification. Although in many cases the aim of a gene transfer experiment is to express the introduced genetic construct (or *transgene*) in the recipient cells and provide those cells with a *gain of function*, gene transfer can also be used to disrupt or inactivate particular endogenous genes (resulting in a *loss of function*), or to perform specific genome modifications. The applications of gene transfer are manifold, and range from the use of mammalian and insect cell cultures for the large-scale commercial production of recombinant antibodies and vaccines, to the transfer of DNA into human patients for the correction or prevention of disease, a field known as *gene medicine* or *gene therapy*. Gene transfer to animal cells is also the first step in the creation of genetically modified whole animals, in which every cell or a specific target population of cells carries a particular alteration. Such animals are used to study gene function and expression, model human diseases, produce recombinant proteins in their milk and other fluids, and to improve the quality of livestock herds and other domestic species. In the last 5 years, animal gene transfer experiments have been conceived on an ever larger scale in an attempt to determine the functions of the many genes discovered in the genome projects (*functional genomics*). Examples of such experiments include systematic DNA-mediated mutagenesis and gene trap programs in the mouse and in the fruit fly, *Drosophila melanogaster*, genome-wide RNA interference experiments in the nematode, *Caenorhabditis elegans*, and novel protein interaction screens based on the yeast two-hybrid system but performed using mammalian cells.

1.2   Historical perspective and some definitions

The concept of gene transfer between cells was first demonstrated in bacteria, which are capable of at least four natural forms of genetic exchange. The first mechanism was discovered in 1928 by Frederick Griffith while working with two different strains of the bacterium *Streptococcus pneumoniae*. He found that the harmless R strain, characterized by rough-looking colonies, could be converted into the virulent S strain, characterized by smooth colonies, if living R cells were mixed with killed S cells before injecting them into mice (*Figure 1.1*). When this experiment was carried out, the

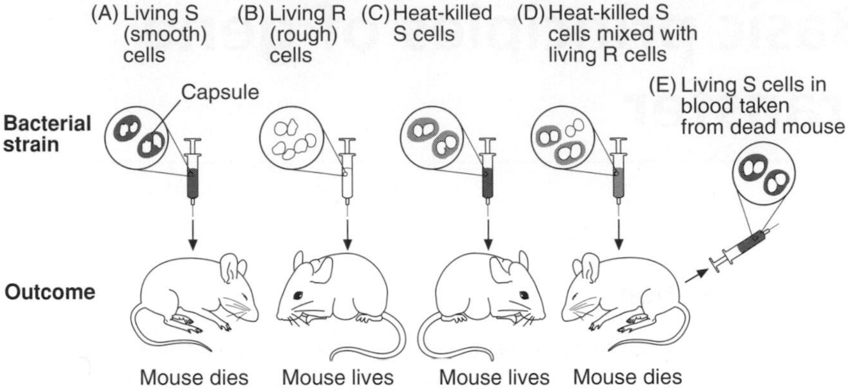

Figure 1.1

Griffith's experiment in 1928 was the first to demonstrate bacterial transformation, but it was over 15 years before the transforming principle was identified as DNA.

mice got sick and died, and living S cells could be recovered from the bodies. Griffith proposed that the genetic instructions to make virulent smooth colonies had somehow been transferred from the killed S cells to the living R cells, and he named this phenomenon *transformation*. However, he was unable to determine the nature of the transforming principle, and it was not until 1944 that Oswald Avery repeated the experiment and established that the substance transferred between cells was DNA. Later, it was discovered that the genetic information present in the S strain but absent from the R strain was a series of genes required to synthesize a proteinaceous capsule that helped the bacteria evade the host's immune response and also gave the colonies their smooth appearance.

A second form of gene transfer known as *conjugation* was discovered by Joshua Lederberg and Edward Tatum in 1946 while working with *Escherichia coli*. This process involved the transfer of DNA between bacterial cells following the establishment of a direct link between them, and in *E. coli* this conduit between cells took the form of a proteinaceous tube known as a pilus. The ability of the cells to construct the pilus and pass DNA through it was encoded on a large plasmid known as the F (for fertility) factor. In most cases, the act of conjugation involved transfer of the plasmid alone, which became established in the recipient cell thereby converting it from an F⁻ to an F⁺ phenotype (*Figure 1.2*). In some cases, however, the F plasmid could integrate into the bacterial chromosome, and conjugation could result in the transfer of chromosomal genes. This process, which was used to construct the first genetic map of *E. coli*, was termed *sexduction* (*Figure 1.3*). Conjugation was soon found to occur in many species of bacteria, and often occurs *between* different species. In some cases it even occurs between bacteria and eukaryotes (see Chapter 3).

It was while studying gene transfer in *Salmonella* that Joshua Lederberg and Norton Zinder discovered in 1951 a new type of gene transfer mediated by bacteriophage. They found that newly formed phage could

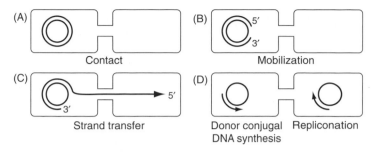

Figure 1.2

The four steps involved in gene transfer by conjugation. (A) Establishment of contact between cells (in *E. coli* and many other bacteria this is achieved by the formation of a conjugal tube known as a pilus). (B) Mobilization (nicking of the donor plasmid to initiate transfer). (C) Strand transfer into the recipient cell. (D) Donor conjugal DNA synthesis (replacement of the donated strand in the donor cell) and repliconation (synthesis of the complementary strand in the recipient cell).

occasionally package some of the host cell's DNA and then transfer it to a second host cell in a subsequent infection. This process was named *trans-duction*. Two forms of transduction could be distinguished – *generalized transduction*, where the phage head was mistakenly stuffed completely with host cell DNA (*Figure 1.4*), and *specialized transduction*, where the phage genome integrated into the bacterial chromosome and became linked to host DNA, and both types of DNA were packaged together. Specialized transduction in *E. coli* infected with bacteriophage λ is shown in *Figure 1.5*. The fourth mechanism of gene transfer in bacteria is mediated by complete cell fusion, and occurs in several genera of bacteria including *Bacillus* and *Streptomyces*.

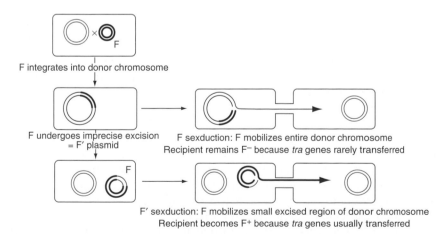

Figure 1.3

Sexduction: the transfer of chromosomal DNA during bacterial conjugation. The F plasmid can mediate sexduction either by conducting the chromosome into which it has integrated, or by exising imprecisely and conducting the chromosomal genes it has captured.

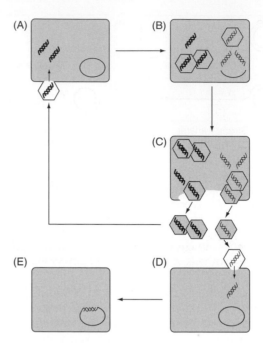

Figure 1.4

Generalized transduction in bacteria occurs when host DNA is accidentally packaged in the phage head and transferred to another cell in the subsequent round of infection. (A) Page infects a bacterial cell and injects its own DNA (thick lines). (B) As DNA is packaged into progeny phage, the bacterial genome breaks up and some is packaged into phage heads. (C) Phage are released from the cell. Those carrying phage DNA initiate new infections, while those carrying bacterial DNA can (D) transfer this to a new host. (E) In the new host, the transduced DNA can integrate or recombine with the host genome, changing the cell's phenotype.

The earliest experiments involving gene transfer to animal cells were reported shortly after the discovery of transduction in bacteria. Perhaps the first significant development was the study by Manker and colleagues in 1956 showing that embryonic chicken cells formed discrete clumps or *foci* when exposed to the genetic material of Rous sarcoma virus, a retrovirus that causes tumors in susceptible birds. The ability of this virus to induce tumors in birds and promote clumping in cultured cells reflects the presence of an *oncogene* (a gene that stimulates cell proliferation) within the viral genome. Under natural conditions, the viral capsid is needed to introduce the oncogene and the rest of the genome into an animal host cell. The unusual process of gene transfer without the viral capsid was named *transfection* to distinguish it from normal infection. A few years later, Ito and colleagues showed that tumors could be induced in rabbits injected with papillomavirus DNA, providing the first evidence for in vivo gene transfer in animals.

In the late 1960s and early 1970s, as the recombinant DNA era began, it became desirable to introduce exogenous DNA into cells artificially, usually

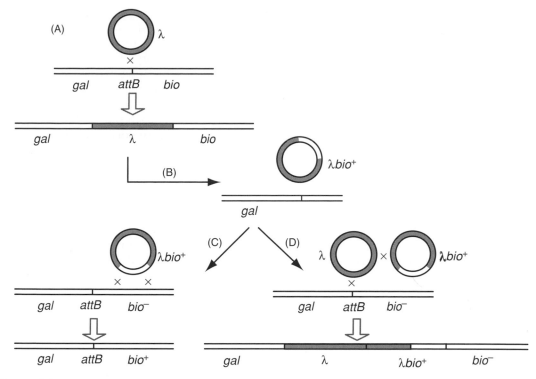

Figure 1.5

Specialized transduction in bacteriophage λ. (A) λ integrates at *attB* between the *gal* and *bio* loci of *E. coli*. (B) Aberrant excision generates a specialized transducing particle λ *bio*⁺ which carries the *bio* gene. Subsequent infection of *bio*⁻ host can lead to (C) replacement transduction by recombination, or (D) addition transduction by integration, the latter generating a λ:λ *bio*⁺ double lysogen which generates high-frequency transduction lysates. Similar events can occur which involve the *gal* locus.

in the form of a plasmid. The reason for this was simple: it was the most efficient way to deliver specific genes and study their effects; it allowed the use of very small plasmids, which were easy to manipulate in vitro. In bacterial genetics, the term *transformation* continued to be used to describe the uptake of naked plasmid or genomic DNA (essentially any DNA which had the potential to transform the phenotype of the recipient cell) while *transfection* was used specifically to describe the uptake of naked phage DNA (or RNA), i.e. nucleic acid which had the potential to initiate a phage replication cycle. For researchers working with animal cells, however, the term *transformation* had never been used in the gene transfer context and was already widely used to describe the cellular phenomenon discussed earlier, i.e. the conversion of normally growing cells to those with an oncogenic phenotype (growth transformation or oncogenic transformation). Therefore, the term *transfection* became generally accepted to mean the introduction of any sort of DNA – phage, plasmid, genomic or otherwise – into an animal cell, in the absence of a biological vector. The same term is also used when analogous methods are used to introduce RNA into cells. The transfection of animal cells thus encompasses all chemical and physical means of nucleic acid delivery, as discussed in Chapter 2. Gene transfer to animal cells mediated by viruses is termed *transduction*, as it is in bacteria, and is the subject of Chapter 4. Chapter 3 considers a further category of gene transfer mechanism which uses bacterial vectors, and is sometimes termed *bactofection*. This can sometimes be analogous to bacterial conjugation, as in the case of HeLa cell transformation by *Agrobacterium tumefaciens*, but in most cases involves invasion of the animal host cell by the bacterium followed by the release of plasmid DNA, which is perhaps more analogous to bacterial cell fusion. The gene transfer methods used with animal cells are summarized in *Table 1.1*. (See references (1–5).)

1.3 Stages of gene transfer and the fate of exogenous nucleic acid

Regardless of the delivery method, gene transfer into animal cells must accomplish three distinct goals (*Figure 1.6*). First, the exogenous genetic

Table 1.1 Overview of gene transfer methods used with animal cells

Method	Advantages	Disadvantages
Transfection (chemical delivery methods)	Highly effective with cultured cells No limitations on transgene size Relatively simple Rapid and suitable for high throughput experiments	Limited in terms of clinical applications Challenge to prepare consistent formulations Some formulations are expensive Cell type dependency
Transfection (physical delivery methods)	High efficiency gene transfer No limitations on transgene size No cell type dependency Suitable for clinical applications	Low throughout Require specific instruments
Transduction (viral vectors)	High efficiency gene transfer Systemic in whole animals Cell-specific targeting possible	Complex cloning required More expensive than transfection methods Safety concerns regarding production of infectious viruses in humans May provoke immune response in mammals
Bactofection (bacterial vectors)	High-efficiency gene transfer	Safety concerns regarding production of infectious bacteria in humans May provoke immune response in mammals Not widely tested

material must be *transported across the cell membrane*. Such transfer is independent of the nature of the genetic material, which is inert and passive at this stage. In physical transfection methods, transport across the membrane is achieved by direct transfer, e.g. in microinjection or particle bombardment where the membrane is breached during delivery, or in electroporation where transient holes are formed, through which DNA and RNA can diffuse. In other delivery methods, the nucleic acid must form some sort of complex which binds to the cell surface before internalization. For example, in chemical transfection methods the complex is formed between nucleic acid and a synthetic compound, while in transduction methods the complex comprises nucleic acid packaged inside a viral capsid.

Once across the cell membrane, the genetic material must be released in the cell and transported to its site of expression or activity. Again, the nucleic acid is passive at this stage. In most transfection methods, DNA or RNA complexes are deposited in the cytoplasm, either directly under the plasma membrane or deeper in the cytosol following escape from the endosomal vesicle. DNA must be transported to the nucleus, a process which depends on the cell's poorly understood intrinsic DNA trafficking system, while RNA can function directly in the cytoplasm. In methods such as particle bombardment and microinjection, it is possible to deliver DNA directly into the nucleus, so intrinsic transport pathways are not required. Many viruses also deliver their nucleic acid cargo to the nucleus as part of the

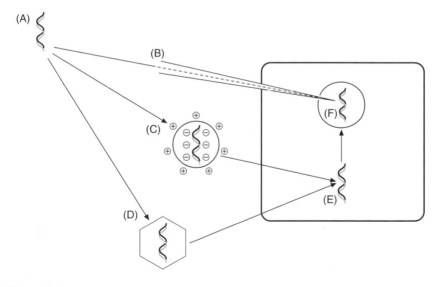

Figure 1.6

Summary of gene transfer mechanisms for introducing genetic material into animal cells. (A) DNA can be (B) introduced directly into the cell, and in some cases into the nucleus, by physical transfection methods. Alternatively (C) the DNA can form a chemical complex with a transfection reagent, resulting in its uptake into the cell, or may (D) be encapsulated within a viral particle (or a bacterial cell). The exogenous DNA first enters the cytoplasm (E) and then the nucleus (F).

infection cycle, often after interaction with cell surface receptors and either internalization within endosomes or direct fusion with the plasma membrane. However, notable exceptions include poxviruses (e.g. Vaccinia virus) and alphaviruses (e.g. Sindbis virus) which replicate in the cytoplasm.

In the final stage of gene transfer, the exogenous genetic material must be *activated*, i.e. released from its complex and rendered competent for expression and/or interaction with the host genome. While exogenous RNA tends to exist only transiently in the host[1], exogenous DNA may be maintained transiently or permanently. Transient expression may be accomplished by transfection with non-replicative DNA, in which case the DNA is diluted and generally lost from the population of cells over a few hours. For plasmids containing a functional viral origin of replication, episomal (extrachromosomal) maintenance may be possible in a compatible cell line, and this may be transient or stable depending on the nature of the replicon and its replication rate (Chapter 2). For example, plasmids with SV40 origins will replicate quickly in simian cells and kill them after a few days, resulting in transient expression. In contrast, plasmids with Epstein–Barr virus latent origins can persist for months in a variety of mammalian cell lines, and plasmids with herpes simplex virus origins can remain episomally stable in neurons for years. In the case of non-replicative DNA, stable expression can only be achieved by integration into the host cell's genome. Some viral transgenes also persist through integration (e.g. retroviruses, adeno-associated virus). In stable transfection experiments with non-replicating DNA, many cells are initially transiently transfected, but the DNA is soon diluted and destroyed. Selection (see Section 1.5) is used to identify the rare cells where integration has taken place, and by this time, the 'transient' DNA is no longer present. (See references (6,7).)

1.4 Monitoring gene transfer: reporter genes

Researchers working with cell lines often have a large number of alternative gene transfer procedures that can be used in their system, and it is generally necessary to compare these procedures and determine their relative efficiencies. It is also desirable in many cases to compare the activities of a series of related regulatory elements in a similar genetic background, such as a transfected cell line or a transgenic animal. When foreign genes are introduced into animal cells, it is possible to follow their expression directly using methods such as RT-PCR, northern blot hybridization, nuclease protection or in situ hybridization, or to study the abundance and distribution of the gene product by immunological assay or functional testing. This can be a laborious approach, however, especially when multiple genes are transferred, and in many cases such methods cannot distinguish between a transgene and a related endogenous gene, or a series of related transgenes.

To avoid these problems, it is conventional to use *reporter genes*, which are marker genes that encode a product that can be detected and, ideally, quan-

---

1 Retrovirus RNA is an exception here, because it is delivered to the nucleus and converted into a DNA copy.

tified using a simple, inexpensive and rapid assay. To study gene transfer efficiency in cells, the reporter gene is linked to a strong and constitutive promoter for the cell system being used in the experiment. For example, in mammalian cells the SV40 early promoter, Rous sarcoma virus LTR promoter and human cytomegalovirus immediate early promoter are generally very active, and are widely used in expression vectors for high-level transgene expression (see Chapter 7). The most commonly used reporter genes in animal cells are *cat*, *lacZ* (or sometimes *gusA*), *luc*, *gfp* and its derivatives and *SEAP*, which are described in more detail in *Table 1.2*. The reporter gene products differ greatly in their half-lives, which is an important consideration for researchers wishing to study rapid inductive responses as opposed

Table 1.2 Properties of reporter genes used with animal cells

Reporter gene (product)	Comments
lacZ (β-galactosidase)	*Source:* E. coli *Activity:* Catalyzes the hydrolysis of β-galactosides, e.g. lactose and many specialized derivative substrates for different assay formats. *Assays:* β-galactosidase assays are nonisotopic. *In vitro* assay formats based on colorimetric (with substrate ONPG) or fluorometric (with substrate MUG) detection systems lack sensitivity, but chemiluminescent systems using 1,2-dioxetane derivatives are highly sensitive. High-resolution *in vivo* histological assays are also available. The chromogenic substrate X-gal can be used for fixed tissue, and the fluorescent substrate FDG can be used for live cells. *Advantages:* Versatile, sensitive and many assay formats available. *Disadvantages:* Some mammalian cells have high endogenous β-galactosidase activity. *Other comments:* β-galactosidase is a stable protein.
gusA (β-glucuronidase, GUS)	*Source:* E. coli *Activity:* Catalyzes hydrolysis of β-glucuronides. *Assays:* Variety of colorimetric, fluorometric and chemiluminescent assay formats, including high-resolution *in vivo* histological assays using the chromogenic substrate X-gluc. Can be used for *in vitro* and *in vivo* non-destructive assays. *Advantages:* Versatile, sensitive and many assay formats available. Smaller gene than *lacZ*. *Disadvantages:* Some mammalian cells have high endogenous β-glucuronidase activity. Other comments: More widely used in plants than in animals due to very low background activity in plant cells. Has been used for tissue specific analysis in C. *elegans*.
cat (chloramphenicol acetyltransferase)	*Source:* E. coli transposon Tn*9* *Activity:* Catalyzes the transfer of acetyl groups from acetyl coenzyme A to chloramphenicol. *Assays:* In vitro assays are isotopic, involving chromatographic separation of acetylated and non-acetylated forms of ^{14}C-chloramphenicol. Such assays have low sensitivity and are expensive, but more recently developed immunological and fluorometric assays are better. *In vivo* CAT assays rarely used due to low resolution. *Advantages:* Minimal background activity in mammalian cells. *Disadvantages:* Low sensitivity, expense, reliance on isotopic assay format. *Other comments:* CAT is a highly stable protein.

Table 1.2 Continued

Reporter gene (product)	Comments
SEAP (secreted alkaline phosphatase)	*Source*: Truncated form of human *PLAP* *Activity*: Removes phosphate groups from a variety of substrates. *Assays*: Nonisotopic, sensitive *in vitro* assays using either colorimetric, fluorometric or chemiluminescent formats to detect secreted protein. Not used for *in vivo* assays. *Advantages*: Secreted protein can be assayed in growth medium without lysing cells, allowing multiple assays for the same culture and further manipulation of cells following assay. *Disadvantages*: High endogenous levels of alkaline phosphatase in some mammalian cells (although SEAP is heat-tolerant allowing endogenous enzyme to be inactivated by heat treatment). Reporter system depends on correct function of the secretory pathway.
luc (firefly luciferase)	*Source*: The firefly *Photinus pyralis* *Activity*: Light produced in the presence of luciferase, its substrate luciferin, oxygen, magnesium ions and ATP. *Assays*: Non-isotopic bioluminescent assays are used *in vitro* and *in vivo*. These are highly sensitive and can be performed in live cells, using lipophilic luciferin esters. *Advantages*: Sensitive, minimal background activity in mammalian cells. *Disadvantages*: Requires expensive detection equipment, some assay formats have limited reproducibility. *Other comments*: Luciferase has a high turnover rate and is thus useful for the study of inducible systems.
GFP (green fluorescent protein) and related proteins	*Source*: Originally the jellyfish *Aequorea victoria*; many fluorescent proteins also isolated from different coral reef species. *Activity*: Intrinsic ability to fluoresce under blue or UV light. *Assays*: Non-isotopic. Used for *in vivo* assays in live cells and animals. Allows monitoring of changes of gene expression in real time, and fusion GFPs allow protein-sorting events to be followed. *Advantages*: Intrinsic activity (no substrate requirements), sensitivity, use in live organisms, many different colors available. *Disadvantages*: The signal from *A. victoria* GFP is not intense enough for some systems. *Other comments*: Improved fluorescent proteins with stronger emission, and emission at different wavelengths, have been generated by mutation allowing multiple studies in a single cell (e.g. GFP, CFP and YFP – green, cyan and yellow). Other colors of fluorescent protein isolated from coral reefs include DsRed, AmCyan and ZsYelow (Clontech). GFP fluorescence is stable and long lasting, but destabilized versions also available.

Abbreviations: ATP, adenosine 5′-triphosphate; CAT, chloramphenicol acetyltransferase; FDG, fluorescein di-β-D-galactopyranoside; G/C/YFP, green/cyan/yellow fluorescent protein; MUG, 4-methylumbelliferyl-β-D-galactoside; ONPG, o-nitrophenyl-β-D-galactopyranoside; PLAP, human placental alkaline phosphatase; SEAP, secreted alkaline phosphatase; UV, ultraviolet; X-gal, 5-bromo-4-chloro-3-indolyl-β-D-galactopyranoside. From Twyman and Whitelaw in The Encyclopedia of Cell Technology, Spier, R.E., Copyright (2000), Reprinted with permission of John Wiley & Sons, Inc.

to longer-term effects. General properties that make an ideal reporter gene product include the availability of quantitative assays that are sensitive over a broad linear range, multiple assay formats, minimal background activity in animal cells, and low toxicity so that the host cell is not affected.

As well as being used to test the efficiency of gene transfer, reporter genes are also used to test the activity of regulatory elements. This is achieved by inserting the putative elements upstream of the reporter gene in specialized *regulatory probe vectors* (*Box 1.1*). The recombinant vectors are then introduced into one or more types of animal cell and quantitative assays

BOX 1.1

Regulatory probe vectors

There are several different types of regulatory probe vectors which are used for the analysis of different types of *cis*-acting element, but each carries a reporter gene such as *cat* or *lacZ* and allows the insertion of putative regulatory elements either upstream or downstream as appropriate. *Promoter probe vectors* are used for the identification of promoters, and carry a reporter gene with an initiation codon but no upstream regulatory sequences. The reporter is only expressed if the inserted regulatory sequence contains a promoter which is active in the cell type used for transfection. Enhancer probe vectors carry a reporter gene driven by a minimal promoter, and are activated only if the inserted regulatory sequence contains an enhancer. There are also vectors available for the analysis of transcriptional termination sites, polyadenylation sites and splice signals (*Figure 1.7*).

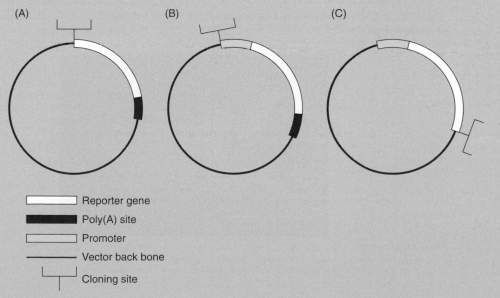

Reporter gene
Poly(A) site
Promoter
Vector back bone
Cloning site

Figure 1.7

General structure of three classes of reporter probe vector. (A) Promoter probe. (B) Enhancer probe. (C) Poly(A) probe.

for the reporter molecule allow the experimenter to gauge indirectly the level of transcriptional activity conferred by the regulatory elements (a second reporter gene under the control of a constitutive promoter is often used to correct for differences in transfection efficiency between cell lines). In the same way, inducible regulatory elements can be tested for their response to particular stimuli by measuring the induction of reporter gene expression when different reagents are added to the culture medium. Reporter genes are also useful in transgenic animals. In many cases, a reporter gene is used simply to confirm transformation – for example, the *Drosophila* genes *white*, *rosy* and *vermillion*, all of which restore wild-type eye color in the appropriate mutant genetic background, are used to confirm transgene integration, as are the *rough* gene, which restores normal eye morphology in a *rough* mutant background, and the *yellow* gene which restores normal body pigmentation. Mouse coat color markers can be used in a similar fashion (p. 145) and the enhanced gene for GFP has been used under the control of an opsin promoter to confirm gene transfer in *Xenopus* by giving the transgenic frogs green fluorescent eyes (*Figure 1.8*)! Regulatory elements can be assayed in transgenic animals using the reporter-transgenic strategy, in which a reporter gene such as *lacZ* is attached to a series of related regulatory elements introduced into different transgenic lines. The availability of histochemical assays for *lacZ* and other reporters allows the reporter gene expression profile and that of the native gene to be compared in the same tissue sections or whole-mount specimens. A common strategy, in both cell lines and animals, is to generate a series of mutated versions of a given reporter construct to narrow down particular *cis*-acting motifs in regulatory elements (an example is considered in Chapter 10, p. 232). Upstream events in the control of gene expression can also be studied, e.g. by addressing the response of reporter genes to the activation or perturbation of specific signaling pathways. See references (7–10).

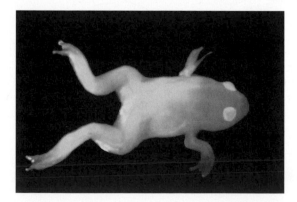

Figure 1.8

Plucky, a transgenic albino African clawed toad (*Xenopus laevis*) showing expression of the enhanced green fluorescent protein in rod cells of the eye, resulting from *gfp* transgene expression under the control of the XOP (*Xenopus* opsin) promoter. Image supplied by Drs David Papermaster, Orson Moritz and Beatrice Tam during research supported by The National Institutes of Health, National Eye Institute and the Foundation Fighting Blindness. Copyright Dr David Papermaster, University of Connecticut Health Center, used with permission.

1.5   Selectable markers for the stable transformation of animal cells

Stable transformation is a very inefficient process when integration of exogenous DNA into the genome is required. Under some circumstances, the foreign DNA introduced into a cell may confer a phenotype that can itself be selected or used as a visible assay for stable transformation. This is unusual, however, so a *selectable marker gene* is normally introduced along with the non-selectable foreign DNA to allow transformed cells to be propagated under conditions in which the high background of non-transformed cells will die or fail to prosper. In the case of integrating viral vectors such as retroviruses, it is conventional to place both the non-selected transgene and the selectable marker in the same vector. Initially, the same strategy was followed for plasmid vectors – the selectable marker was generally included on the same vector as the non-selectable foreign DNA, so that the two genes would co-integrate and selection for the marker would necessarily identify cells carrying the non-selectable gene of interest. However, it was shown that two discrete plasmids cotransfected into mammalian cells also resulted in a high frequency of co-integration, i.e. where the initially unlinked genes were integrated together, usually at the same locus. Typically, a 10:1 ratio of non-selectable to selectable vector is used for cotransformation, so that cells transformed with the marker are highly likely to be cotransformed with the non-selectable DNA. The markers are used to select founder cells that give rise to stably transformed cell lines. Cells are allowed to grow under the appropriate selective regime for about ten generations. At this point, individual clones of surviving cells are isolated and used to found new lines.

The first selectable marker to be widely used with animal cells was *Tk*, encoding the enzyme thymidine kinase (TK) which is required to synthesize the nucleotide dTTP from thymidine. This is an *endogenous marker*, i.e. it represents an activity already present in most mammalian cells, and therefore requires a *Tk⁻* cell background for positive selection (selection for the *presence* of the marker and the property it confers on the cell). Like most endogenous markers, *Tk* encodes a non-essential enzyme involved in nucleotide biosynthesis. Such enzymes are non-essential because there are two alternate nucleotide biosynthesis pathways in mammalian cells: under standard growth conditions, TK is not required because dTTP can be synthesized afresh from carbamyl phosphate (the de novo pathway). However, if cells are grown in medium containing the inhibitor aminopterin, the de novo pathway is blocked and dTTP must be synthesized from thymidine (the salvage pathway), which requires TK. Other endogenous markers are listed in *Table 1.3*, and their role in nucleotide biosynthesis is summarized in *Figure 1.9*. The problem with such markers is that the number of cell types available for transfection is limited to those where appropriate mutant lines have been developed. There has therefore been much interest in the development of novel markers that can be used in all mammalian cells.

Such *dominant selectable markers* (dominant because they are typically derived from a bacterial source for which there is no competing activity in any animal cell) are often drug resistance genes. Cells can be propagated in

Table 1.3 Endogenous markers used for selection in mammalian cells. Most of the markers encode enzymes involved in the nucleotide biosynthesis salvage pathway, and can therefore be used both for positive selection (selection for the presence of the marker) and negative selection (selection for the absence of the marker). Positive selection is achieved using salvage pathway mutant cell lines in combination with drugs that block the de novo nucleotide biosynthesis pathway. Negative selection is achieved using toxic base analogs that are incorporated into DNA only if the salvage pathway is active

Marker	Product and function	Principles of selection
Ada	Adenosine deaminase; converts adenosine to inosine	Xyl-A (9-β-D-xylofuranosyl adenosine) is an adenosine analog, which is toxic if incorporated into DNA. ADA detoxifies Xyl-A added to the growth medium by converting it to Xyl-I, its inosine equivalent. There is a low background of ADA activity in most mammalian cells and 2'-deoxycoformycin, an ADA inhibitor, is therefore included in the selection medium.
Aprt	Adenine phosphoribosyl-transferase; converts adenine to AMP	Positive selection: Adenine, and azaserine, to block *de novo* dATP synthesis, so only cells using salvage pathway survive. Negative selection: Toxic adenine analogs (e.g. 2,6-diaminopurine), which are incorporated into DNA only in cells with APRT activity.
Cad	Carbamyl phosphate synthase; aspartate transcarbamylase; dihydro-oroatase.	These are the first three steps in *de novo* uridine biosynthesis. Positive selection: PALA (*N*-phosphonacetyl-L-aspartate) inhibits the aspartate transcarbamylase activity of CAD.
Dhfr	Dihydrofolate reductase; converts folate to dihydrofolate and then to tetrahydrofolate	Positive selection: DHFR is required for several reactions in *de novo* and salvage nucleotide/amino acid biosynthesis, hence selection is carried out in nucleotide-free medium. However, *Dhfr* is typically used as an amplifiable marker with the inhibitor methotrexate (see *Table 1.5*).
Hgprt	Hypoxanthine-guanine phosphoribosyltransferase; converts hypoxanthine to IMP and guanine to GMP	Positive selection: Hypoxanthine and aminopterin, to block *de novo* IMP synthesis, so only cells using salvage pathway survive. Selected on HAT medium, which contains hypoxanthine, aminopterin and thymidine. Negative selection: Toxic guanine analogs (e.g. azaguanine) which are incorporated into DNA only if there is HGPRT activity in the cell.
Tk	Thymidine kinase, converts thymidine to dTMP	Positive selection: Thymidine, and aminopterin to block *de novo* dTTP synthesis, so only cells using salvage pathway survive. Selected on HAT medium (see above). Negative selection: Toxic thymidine analogs (e.g. 5-bromo-deoxyuridine, ganciclovir, FIAU) which are incorporated into DNA only if there is TK activity in the cell.

Abbreviations: AMP, adenosine monophosphate; HAT medium, hypoxanthine, aminopterin, thymidine; IMP, inosine monophosphate; GMP, guanosine monophosphate; PALA, (*N*-phosphonacetyl-L-aspartate); dTMP/dTTP deoxythymidine mono/tri phosphate; Xyl-A, (9-β-D-xylofuranosyl adenosine); Xyl-I, (9-β-D-xylofuranosyl inosine). From Twyman and Whitelaw in The Encyclopedia of Cell Technology 2nd Ed; Spier, R.E., Copyright (2000), Reprinted with permission of John Wiley & Sons, Inc.

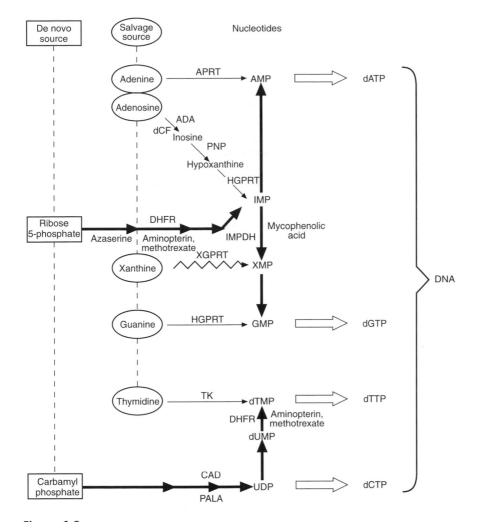

Figure 1.9

Nucleotide biosynthesis and commonly used endogenous selectable markers. *De novo* nucleotide synthesis begins with the boxed precursors and follows the thick arrows. In the absence of *de novo* substrates, or where the *de novo* pathway is blocked, the cell can use pyrimidine and purine salvage pathways, beginning with the circled precursors and following the thin arrows. Mammalian cells cannot convert xanthine to XMP, but this reaction is carried out by the *E. coli* enzyme XGPRT, encoded by the *gpt* gene, allowing this to be used as a dominant selectable marker in mammalian cells (zig-zag arrow). Eventually, the nucleotides are converted into dNTPs (open arrows) and incorporated into DNA. The salvage enzymes are not required for cell growth when *de novo* substrates are available so salvage pathway mutants are viable under normal growth conditions. However, if the *de novo* pathway is blocked using any of the inhibitors (shown in bold), genes encoding salvage pathway enzymes can be used as selectable markers when salvage precursors are also provided. For example, *Tk* can be used to circumvent aminopterin-inhibited *de novo* dTMP synthesis if the salvage precursor, thymidine, is included in the growth medium. Alternatively, *de novo* and salvage inhibitors can be overcome by amplifiable selection, e.g. PALA, an inhibitor of *de novo* UDP synthesis, can be overcome by amplification of the *Cad* gene, and methotrexate, an inhibitor of several *de novo* and salvage reactions, can be overcome by amplification of the *Dhfr* gene. Salvage pathway markers can also be counterselected using toxic base or nucleoside analogs. These are toxic when incorporated into DNA via the salvage pathway, but innocuous if nucleotides are synthesized *de novo*. For example, 5-bromodeoxyuridine (bdUr) is a thymidine analog incorporated into DNA following conversion into a nucleotide by TK. In the absence of TK, bdUr is not phosphorylated and is not incorporated into DNA. Azaserine blocks several *de novo* reactions, the most important of which is shown. Abbreviations of enzymes and inhibitors: APRT, adenine phosphoribosyltransferase; ADA, adenosine deaminase; CAD, carbamyl phosphate synthase/aspartate transcarbamoylase/dihydroorotase; dCF, deoxycoformycin; DHFR, dihydrofolate reaductase; HGPRT, hypoxanthine-guanine phosphoribosyltransferase; IMPDH, inosine monophosphate dehydrogenase; PALA, *N*-phosphonacetyl-L-aspartate; PNP, purine nucleoside phosphorylase; TK, thymidine kinase; XGPRT, xanthine-guanine phosphoribosyltransferase. From Twyman and Whitelaw in The Encyclopedia of Cell Technology, 2nd Ed, Spier, R.E., Copyright (2000), Reprinted with permission of John Wiley & Sons, Inc.

Table 1.4 Dominant selectable markers used in animal cells. These markers confer a drug-resistance phenotype upon the host cell which cannot be replaced by any endogenous activity

Marker	Product (and source species)	Principles of selection
AS	Asparagine synthase (*E. coli*)	Bacterial enzyme uses ammonia as amide donor unlike mammalian equivalent. Hence cells transformed with *AS* grow on asparagine-free medium containing the toxic glutamine analog albizziin.
ble	Glycopeptide-binding protein (*Streptoalloteichus hindustanlus*)	Confers resistance to glycopeptide antibiotics bleomycin and pheomycin (and its derivative, Zeocin™).
gpt	Guanine-xanthine phosphoribosyltransferase (*E. coli*)	Analogous to *Hgprt* in mammals, but possesses additional xanthine phosphoribosyltransferase activity allowing survival in medium containing aminopterin and mycophenolic acid (*Figure 1.9*).
hisD	Histidinol dehydrogenase (*Salmonella typhimurium*)	Confers resistance to histidinol.
hpt	Hygromycin phospho-transferase (*E. coli*)	Confers resistance to hygromycin-B.
neo	Neomycin phosphotransferase (*E. coli*)	Confers resistance to aminoglycoside antibiotics (e.g. neomycin, kanamycin, G418).
pac	Puromycin N-acetyltransferase (*Streptomyces alboniger*)	Confers resistance to puromycin.
trpB	Tryptophan synthesis (*E. coli*)	Confers resistance to indole.

From Twyman and Whitelaw in The Encyclopedia of Cell Technology, 2nd Ed, Spier, R.E., Copyright (2000), Reprinted with permission of John Wiley & Sons, Inc.

normal medium and then transformed cells can be selected by adding the drug at the appropriate concentration. Commonly used dominant selectable markers are listed in *Table 1.4*. One disadvantage of these bacterial markers is that the concentration range over which the selective agents are active tends to be narrow. Such markers are therefore not suitable for stepwise selection for increased transgene copy number. As discussed in more detail in Chapter 2, in situ transgene amplification is one way to generate high-yield transformed cell lines, but this requires markers that can be selected in a stepwise manner over a range of selective conditions. The predominant example of this type of *amplifiable marker* is *Dhfr*, which encodes the enzyme dihydrofolate reductase (DHFR). Increasing resistance to methotrexate, a competitive inhibitor of DHFR, usually correlates with increased *Dhfr* gene copy number, and co-amplification of the non-selected donor gene. Amplifiable markers are highly efficient, but *Dhfr* is endogenous and the background of wild-type cells that express and amplify their wild-type *Dhfr* locus can be a problem. For this reason such markers tend to be used in mutant cell lines, which again restricts the number of cell types available for transfection (*Table 1.5*). This problem can be overcome by using a second marker such as *neo*, which allows non-transformed cells to be killed off prior to any amplification regime. (See references (10–14).)

Table 1.5 Markers that can be amplified *in situ* in mammalian cells by increasing the concentration of a competitive inhibitor

Marker	Product	Amplifying selective drug
Ada	Adenosine deaminase	Deoxycoformycin
AS	Asparagine synthase	β-aspartylhydroxamate
Cad	Aspartate transcarbamylase	*N*-phosphonacetyl-L-aspartate
Dhfr	Dihydrofolate reductase	Methotrexate
gpt	Xanthine-guanine phospho-ribosyltransferase	Mycophenolic acid
GS	Glutamine synthase	Methionine sulfoxamine
Hgprt	Hypoxanthine-guanine phosphoriboosyltransferase	Aminopterin
Impdh	Inosine monophosphate dehydrogenase	Mycophenolic acid
Mt-1	Metallothionein 1	Cd^{2+}
*M*res	Multidrug resistance: P-glycoprotein 170 gene	Adriamycin, colchicine, others
Odc	Ornithine decarboxylase	Difluoromethylornithine
Rnr	Ribonucleotide reductase	Hydroxyurea
Tk	Thymidine kinase	Aminopterin
Umps	Uridine monophosphate synthases	Pyrazofurin

From Twyman and Whitelaw in The Encyclopedia of Cell Technology, 2nd Ed, Spier, R.E., Copyright (2000), Reprinted with permission of John Wiley & Sons, Inc.

1.6 Modification of the endogenous genome

Although most gene transfer experiments are designed to incorporate new genetic information into the cell, there are also many different ways in which endogenous genes can be disrupted or inactivated, or modified in more subtle and elegant ways. Occasionally, when foreign DNA is introduced into animal cells with the aim of inserting and expressing a new transgene, the exogenous DNA interrupts an endogenous gene and generates a mutant phenotype. This phenomenon can be exploited to knock out genes deliberately, and if such experiments are carried out on a suitably large scale, then mutant libraries representing the entire genome can be generated by insertional mutagenesis. The great advantage of this strategy is that the interrupted genes carry an insertional DNA tag, which can be used to clone the flanking sequences and isolate the mutated gene, providing a handy short-cut to cloning. Such strategies are discussed in more detail in Chapter 9. Although random DNA integration is the usual outcome of a gene transfer experiment, it is also possible to promote homologous recombination between exogenous and endogenous DNA resulting in the replacement of endogenous DNA sequences with constructs prepared in the laboratory. This type of experiment, which is known as gene targeting, allows very precise manipulation to be carried out, even the introduction of specific point mutations into preselected genes, and therefore facilitates the creation of any mutant cell or organism of interest. Unfortunately, while the random DNA integration that usually results from gene transfer is very rare, genetic exchange by homologous recombination is even rarer, and very powerful selective strategies are often required to identify targeting events. Also, gene targeting has been achieved in only a few animal cell types, so it cannot be applied to a large range of organisms (Chapter 6). In contrast, there are many other strategies for gene inactivation which involve conventional gene transfer and integration, but the transgene expresses a product which specifically interferes with the expression of a target endogenous gene. Such approaches, which include the expression of antisense RNA, ribozymes, short interfering RNAs and antibodies, are discussed in Chapter 9. The following three chapters now explore gene transfer methods in more detail.

References

1. Griffith F (1928) The significance of pneumococcal types. *J Hyg* **27**: 113–159.
2. Lederberg J, Tatum EL (1946) Gene recombination in *E. coli*. *Nature* **158**: 558.
3. Lederberg J, Lederberg EM, Zinder ND, Lively ER (1951) Recombination analysis of bacterial heredity. *CSH Symp Quant Biol* **16**: 413–443.
4. Manker RA, Groupe V (1956) Discrete foci of altered chicken embryo cells associated with Rous sarcoma virus in tissue culture. *Virology* **2**: 838–840.
5. Ito Y (1960) A tumor reducing factor extracted by phenol from papillomatous tissues of cotton tail rabbits. *Virology* **12**: 596–601.
6. Twyman RM, Whitelaw CAB (2000) Animal cell genetic engineering. In: Spier RE (ed.) *Encyclopedia of Cell Technology*. John Wiley & Sons Inc., NY, pp 737–819.
7. Liu D, Chiao EF, Tian H (2004) Chemical methods for DNA delivery: An overview. In: Heiser WC (ed.) *Gene Delivery to Mammalian Cells. Volume 1 – Nonviral Gene Transfer Techniques*. Humana Press, Towata, NJ, pp 3–23.
8. Kain SR, Ganguly S (1995) Uses of fusion genes in mammalian transfection. In: *Current Protocols in Molecular Biology*, Supplement 29. John Wiley & Sons, NY, pp 9.6.1–9.6.12.
9. Walsh S, Kay SA (1997) Reporter gene expression for monitoring gene transfer. *Curr Opin Biotechnol* **8**: 617–622.
10. Sambrook J, Russel DW (2001) Analysis of gene expression in cultured mammalian cells. In: Sambrook J, Russel DW (eds) *Molecular Cloning: A Laboratory Manual*. CSH Laboratory Press, Cold Spring Harbor, NY, pp 17.1–17.97.
11. Chisholm V (1995) High efficiency gene transfer to mammalian cells. In: Glover DM, Hames BD (eds) *DNA Cloning: A Practical Approach. Volume 4, Mammalian Systems* (second edition). IRL Press, Oxford, UK, pp 1–42.
12. Kingston RE (1995) Stable transfer of genes to mammalian cells. In: *Current Protocols in Molecular Biology* (supplement 14). John Wiley & Sons, New York, pp 9.5.1–9.5.6.
13. Kauffman RJ (1990) Selection and coamplification of heterologous genes in mammalian cells. *Methods Enzymol* **185**: 537–566.
14. Bebbington C (1995) Use of vectors based on gene amplification for the expression of cloned genes in mammalian cells. In: Glover DM, Hames BD (eds) *DNA Cloning: A Practical Approach. Volume 4, Mammalian Systems* (second edition). IRL Press, Oxford, UK, pp. 85–112.

Chemical and physical transfection

2

2.1 Introduction

Historically, the first gene transfer protocols to be developed for animal cells did not use biological vectors such as viruses or bacteria to carry the exogenous nucleic acid across the cell membrane. Instead, naked DNA was mixed with particular chemicals to form synthetic complexes, which would either interact with the cell membrane and promote uptake by endocytosis, or fuse with the membrane and deliver the DNA directly into the cytoplasm. Since the late 1960s, such *chemical transfection* methods have been widely used and many different complex-forming agents have been described. One drawback of these methods, however, is that they are generally inefficient for gene transfer in vivo. In contrast, *physical transfection* methods are efficient for both in vitro and in vivo gene transfer. Synthetic complexes are not required in physical transfection methods because the gene transfer process involves breaching the cell membrane and introducing the nucleic acid directly into the cell, and in some cases directly into the nucleus. In simple terms, chemical transfection methods 'trick' the cell into taking up DNA from the surroundings while physical transfection methods use brute force rather than trickery to achieve DNA transfer (*Figure 2.1*). Although

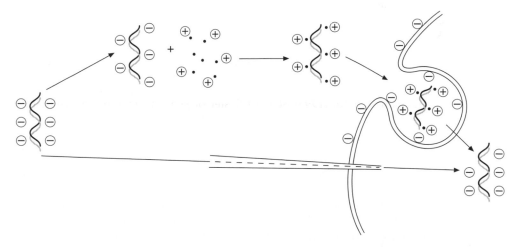

Figure 2.1

Summary of DNA uptake mechanisms of DNA uptake in transfection methods – chemical transfection involves the association between DNA and positively charged molecules allowing the complex to interact with the negatively charged cell membrane. In contrast, physical tranfection involves direct transfer through the plasma membrane.

there are advantages and disadvantages to both sets of procedures, some of the most efficient transfection methods in use today involve a combination of chemical and physical processes, e.g. the Nucleofector technique described below.

Chemical and physical transfection methods were first used for the transfer of naked, wild-type viral DNA into animal cells, but both groups of techniques are now more widely used for the introduction of plasmid vectors and recombinant viral genomes carrying specific transgenes of interest. Since viral vectors are described in detail in Chapter 4, this chapter includes a section describing the principal types of plasmid vector used with transfection protocols. Before considering these vectors, however, we describe the transfection methods themselves.

2.2   Chemical transfection methods

General principles

Chemical transfection methods have to overcome a number of boundaries to deliver active DNA into the nucleus. Unlike physical gene transfer methods, which mechanically breach both the cell membrane and in many cases the nuclear membrane, chemical transfection methods have to persuade the cell to interact with and process exogenous DNA, and eventually deliver at least some intact DNA molecules to the nucleus. As discussed in Chapter 1, the first boundary to gene transfer is the cell membrane, which is hydrophobic and negatively charged. DNA, which is hydrophilic but also negatively charged, can only interact with the cell membrane if it is either enclosed in a fusogenic capsule (as in transfection via liposomes or via cell/protoplast fusion) or sequestered into a complex that carries a net positive charge (as in all other chemical transfection methods). The function of the synthetic complex, or '*transfection reagent*', is therefore either to form a positively charged complex or a fusogenic particle in combination with DNA.

The second boundary to successful transfection, and still the one which causes the most difficulty in chemical transfection methods, is DNA transfer to the nucleus[1]. DNA encapsulated within fusogenic particles is deposited in the cytoplasm under the cell membrane following membrane fusion, and is thought to find its way to the nucleus via an intrinsic transport pathway, although this process is still poorly understood. In contrast, complexes taken up by endocytosis are transported in acidic endosomes, eventually to be deposited in lysosomes and degraded. To achieve a high level of transfection efficiency, the DNA must 'escape' from the endosome into the cytosol and find its way to the nucleus, probably via the same pathway as DNA delivered through the fusogenic route. Certain reagents such as chloroquine can be included in the transfection protocol because they are known to help the DNA escape into the cytoplasm by disrupting endosomes and sabotaging the endosomal transport pathway. A number of

---

1 For transfection with RNA, the boundaries that must be overcome are the plasma membrane and, for RNA delivered via the endosomal pathway, escape into the cytoplasm. Transfer to the nucleus is not required.

peptides with endosome-disrupting properties have also been described. For transfer to the nucleus, the DNA sequence itself can be important. For example, vectors carrying the SV40 virus origin of replication are more efficiently transported to the nucleus than similar vectors lacking this sequence. The inclusion of certain peptides in the transfection mix, including peptides with canonical nuclear localization sequences, can also promote the nuclear import of exogenous DNA.

The final barrier to efficient transfection is activation of the exogenous DNA by dissociation from the complex once within the cell. Although the complex enables DNA to enter the cell and protects it from nucleases, only free DNA is available for expression or interaction with the host cell's genome. The importance of this step was shown in experiments where animal oocyte nuclei were injected with naked DNA and with DNA–lipid complexes: only the former were capable of transgene expression. It is thought that dissociation occurs either by simple diffusion (in the case of calcium phosphate transfection) or, in other transfection methods, when the positively charged complex interacts with negatively charged intracellular lipids and other molecules, and is neutralized. (See references (1,2).)

Calcium phosphate

Transfection using calcium phosphate was arguably the first chemical transfection method to be used with animal cells. The procedure was developed in 1973 by Graham and van der Erb (3) for the introduction of adenovirus DNA into rat cells, eight years after the first description of transfection mediated by DEAE-dextran (see below). However, the presence of calcium was shown to be responsible for the successful transformation of human cells with genomic DNA in a report by Szybalska and Szybalski published in 1962 (4), a report that also pioneered the HAT selection technique (see *Table 1.3*). These authors are rarely credited with their discovery[2]. The first mammalian cell lines stably transfected with plasmid DNA were also produced by calcium phosphate transfection, in 1978 (5).

Calcium phosphate is probably the most widely used transfection method with established cell lines, and this is because it is simple, reliable, applicable to many cultured cell lines, and the reagents are inexpensive. It can be used both for transient and stable transformation. The principle of the technique is that DNA in a buffered phosphate solution is mixed gently with calcium chloride, which causes the formation of a fine DNA–calcium phosphate coprecipitate. The precipitate settles onto the cells and some of the particles are taken up by endocytosis (*Figure 2.2*). The most efficient transfection occurs in cells growing as a monolayer, because these cells become evenly coated with the precipitate. However, not all cells are equally amenable to this transfection method, some are sensitive to the density of the precipitate, and for some reason the efficiency of transfection in primary cells is very poor. As originally described by Graham and van der Erb, the DNA-phosphate solution was buffered with HEPES (*N*-2-hydroxyethylpiperazine-*N'*-2-ethanesulfonic acid) (*Protocol 2.1*) but an

---

2 Elizabeth Szybalska and Waclaw Szybalski also coined the term 'gene therapy', in 1962, but are rarely recognized for this either!

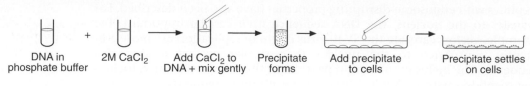

Figure 2.2

Calcium-phosphate-mediated transfection.

improved protocol using BES buffer (*N,N*-bis(2-hydroxyethyl)-2-aminoethanesulfonic acid) can be up to 100 times more efficient for stable transfection when high-quality plasmid DNA is used (*Protocol 2.2*). BES allows a very fine precipitate to form slowly in the culture medium over a period of hours, which probably reduces the cytotoxicity of the procedure and enhances DNA uptake. Perhaps the major disadvantage of this method is that the success of transfection depends on the particle size, which depends on many factors including the way in which the reagents are mixed together. Since this is difficult to standardize, results can be highly variable and difficult to reproduce under some conditions. (See references (3–9).)

Transfection with DEAE-dextran

DEAE-dextran (diethylaminoethyl-dextran) is a soluble polycationic carbohydrate that forms aggregates with DNA through electrostatic interactions. The principle of all transfection methods involving polycationic complexes is that an excess of the transfection reagent, in this case DEAE-dextran, provides the entire complex with a net positive charge, which allows it to interact with the negatively charged cell membrane and promotes uptake by endocytosis (*Figure 2.3*). The complexes are very small compared to the particles formed during calcium phosphate transfection, and one consequence of this is that much less DNA can be used in each transfection experiment. DEAE-dextran was the first transfection reagent to be developed (in 1965) (10) and was very widely used until the advent of lipofection reagents in the 1990s. Like the calcium phosphate method, the reagents are inexpensive and the procedure is simple and efficient, although the efficiency varies according to the cell line. DEAE-dextran-mediated

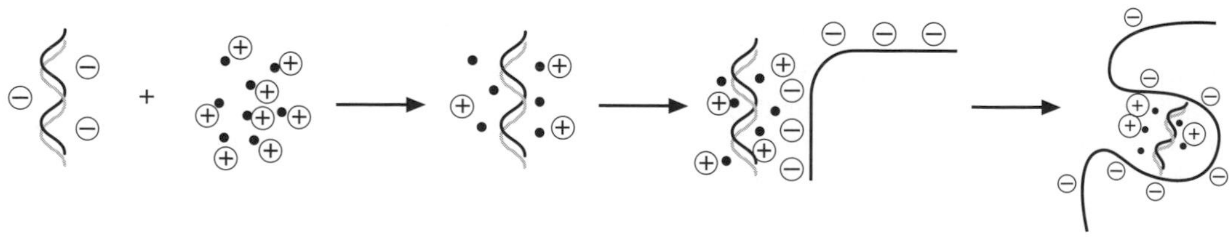

Figure 2.3

Transfection with lipoplexes and polyplexes involves the formation of positively charged complexes which interact with the plasma membrane and stimulate uptake by endocytosis.

transfection is not particularly efficient for the production of stably transformed cell lines, and has fallen out of favor recently mainly for this reason.
However, in the best cases the method can achieve up to 80% transient
transfection efficiency.

The original DEAE-dextran transfection protocol was improved by Lopata
et al. (11) and by Sussman and Milman (12), by the inclusion of so-called
chemical shock treatments to increase the efficiency of DNA uptake. An
osmotic shock following transfection, caused by flooding the cells with a
10–30% solution of glycerol or dimethylsulfoxide (DMSO) increases the
amount of DNA taken up from the surrounding medium, probably by
changing the permeability of the cell membrane to polycationic complexes.
As with other transfection parameters, the optimal concentration and exposure time must be determined empirically. Chemical shocks have also been
used in combination with other transfection methods, including calcium
phosphate, lipofection and electroporation. (See references (10–12).)

Lipofection

Liposomes are hydrophobic, unilaminar phospholipid vesicles into which
DNA can be packaged to form a fusogenic particle. When mixed with cells
in culture, such vesicles can fuse with the cell membrane and deliver DNA
directly into the cytoplasm (*Figure 2.4*), a delivery mechanism similar to that
of cell and protoplast fusion (see Chapter 3). Gene transfer mediated by liposomes was first described by Fengler in 1980 (13). The original liposome
transfection techniques were no more efficient than calcium phosphate
transfection and suffered the further disadvantage that the preparation of
DNA-containing liposomes was complicated and labor-intensive. One
particular advantage of the method, however, was the ability to transform
mouse cells in vivo by injecting liposomes into the tail vein. Indeed liposome-mediated transfection was the first non-viral technique devised
specifically for in vivo DNA transfer, making it suitable for gene therapy.
The efficiency of liposome-mediated gene transfer can be enhanced by incorporating viral proteins that facilitate the active fusion between viral
envelopes and cell membranes. Such fusogenic particles have been termed
virosomes. For example, virosomes have been prepared from the envelopes
of Sendai virus allowing the delivery of DNA to astroglial cultures.

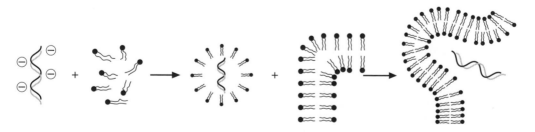

Figure 2.4

Transfection with liposomes is achieved via a fusogenic mechanism, whereby the artificial lipid
vesicle fuses with the plasma membrane and deposits the DNA into the cytoplasm.

A breakthrough in lipid-mediated transfection came with the development of cationic/neutral lipid mixtures, which spontaneously associate with negatively charged DNA to form complexes known as *lipoplexes*. The first cationic lipid compound to be described was DOTMA (*N*-(2,3-dioleyl-oxypropyl) *N, N, N*-trimethylamonium chloride) which forms liposomes when mixed with water but assembles into spontaneous lipoplexes with DNA. Residual positive charge in the complex then interacts with negatively charged molecules on the cell surface, facilitating uptake of the exogenous DNA by endocytosis (*Figure 2.3*). Technically, this is one of the simplest transfection protocols (*Protocol 2.3*). Lipid–DNA complexes are prepared simply by mixing DNA with the lipid preparation in serum-free medium, then the mixture is added to the cells. This facilitates rapid and efficient DNA uptake, and is gentle enough to be applicable to both conventional plasmid vectors and larger constructs such as bacterial and yeast artificial chromosomes (BACs, YACs). Lipofection, as this technique has become known, is highly reproducible and extremely efficient for both transient and stable transfection. It allows up to 90% of cells in culture to be transiently transfected, and demonstrates stable transfection efficiencies up to 20-fold greater than standard chemical transfection methods. One drawback to this approach is that the lipids are usually difficult to prepare in the laboratory. They must therefore be purchased from a commercial source, and they are very expensive.

There are now many different lipid preparations available from different suppliers, varying in price, efficiency and suitability for different cell lines. A selection of lipid reagents is shown in *Table 2.1*. Common structural principles in the most efficient cationic lipid-based transfection reagents include a positively charged head group, a linker, and a hydrophobic 'anchor'. The head group generally contains between one and four amine groups, and this is joined either via a glycerol linkage to an aliphatic hydrocarbon chain anchor, or through a variety of linkages to a cholesterol anchor (16). Transfection can be enhanced by the addition of a neutral 'helper lipid' such as DOPE (dioleoylphosphatidylethanolamine) or cholesterol, although this is not always required. Two important features of the lipoplex that affect transfection efficiency are the size of the complex, which ideally should be greater than 200 nm, and the charge ratio (i.e. the ratio of amine groups in the reagent to phosphate groups in the DNA), which should be slightly greater than 1 for transfection of cultured cells but much higher for in vivo transfection. Lipofection is the most widely used chemical gene transfer method for gene therapy. (See references (13–23).)

Cationic polymers

Unlike the lipid reagents discussed above, cationic polymers are hydrophilic and soluble in aqueous solvents, but like the lipid reagents they form spontaneous complexes with DNA (sometimes termed *polyplexes*). Some relatively simple and linear cationic polymers that have been used include spermine (a naturally occurring polyamine), histones and synthetic poly-L-lysine. One of the first cationic polymers to be used as a transfection reagent was Polybrene (1,5-dimethyl-1,5-diazaundecamethylene polymethobromide) which facilitates the high-efficiency transfection of certain

Table 2.1   A selection of cationic lipid transfection reagents developed for gene transfer (for the sake of brevity, reagents developed specifically for the introduction of oligonucleotides and siRNA into animal cells are not included, nor are reagents developed for individual cell lines or cell types)

Reagent or reagent family	Composition, if known	Supplier
CellFectin	Cationic lipopolyamine	Invitrogen
CLONfectin	Cationic lipid	BD Biosciences/CLONTECH
CytoFectene	Cationic lipid	Amersham
Cytofectin GS	Cationic lipid	Glen Research
DOSPAR	Polycationic lipid	Roche Applied Science
DOTAP	Cationic lipid	Roche Applied Science
Effectegene	Nonliposomal lipid	Qiagen
Escort family	Cationic lipids	Sigma Aldritch
FluoroFectin family	DOTAP, neutral lipid and fluorescent tracer	Qbiogene
FuGENE 6	Nonliposomal lipid	Roche Applied Science
GeneLimo family	Polycationic lipid and helper	CPG
GenePORTER family		Gene Therapy Systems
GeneSHUTTLE family	Polycationic lipids	Qbiogene
Genetransfer	Cationic liposome	Wako
GenFect	Cationic lipid	MoleculA
LipofectAMINE family	Polycationic lipid	Invitrogen
Lipofectin	DOTMA and DOPE	Invitrogen
LipoGen	Nonliposomal lipid	InvivoGen
LipoTaxi	Cationic lipid	Strategene
LipoVec	Cationic phosphonolipid mix	InvivoGen
MaxFect	Cationic lipid	MoleculA
Metafectene	Cationic lipid	Biontex
SureFECTOR	Cationic lipid and neutral lipid	B-Bridge International
Targetfect F1	Cationic lipid	Targeting Systems
Tfx family	Cationic lipids	Promega
TransFAST	Cationic lipid	Promega
Transfectam	Cationic lipids	Promega
TransFectin	Cationic lipid	Bio-Rad
TransIT-Insecta	Cationic lipid	PanVera Corp
TransMessenger		Qiagen
UniFECTOR	Polyationic lipid and neutral lipid	B-Bridge International
X-tremeGENE family		Roche Applied Science

cell lines, including CHO cells, chicken embryo fibroblasts, NIH 3T3 cells and myeloid cells, which can be refractory to transfection by calcium phosphate. For other cell lines, Polybrene has been shown to be no more efficient than calcium phosphate and it has not gained widespread use.

More recently, several new classes of cationic reagents have been developed, including synthetic polyamines, polyethyleneimines (PEIs) and dendrimers. Polyamines, as the name indicates, are organic molecules carrying multiple amine groups. PEIs are hydrocarbon chains with nitrogens at every third position, and these can be linear or branched. The numerous amine groups provide a large number of positive charges for interaction with DNA. Dendrimers are highly complex molecules, built in layers from a central initiator such as ammonia (trivalent) or ethylenediamine (tetravalent) to adopt an overall spherical or star-like geometry with multiple internal and external amine groups. While they have a tendency to be more cytotoxic than lipid transfection reagents, the advantage of dendrimers is that their shape and size can be controlled quite precisely, so the resulting DNA complexes are homogeneous and of a defined size. As is the case for lipid transfection reagents, many proprietary polyamines, PEIs and dendrimers are now available from commercial sources each claiming to maximize transfection efficiency (*Table 2.2*). Factors affecting transfection efficiency include the polymer size and structure, the ratio of polymer to DNA, as well as the cell type and buffer. (See references (23–26).)

Enhancing transfection efficiency

We have already discussed some chemical treatments such as DMSO and glycerol that help to increase transfection efficiency by enhancing the process of DNA uptake. Further reagents are available which increase transfection efficiency in other ways, e.g. by promoting DNA escape from endosomal vesicles, promoting fusion with membranes, directing the transfection complex or exogenous DNA to the nucleus or enhancing transgene expression.

In some mammalian transient transfection systems, for example, cotransfection with a plasmid carrying the adenoviral VA_I gene (see p. 67) helps to increase the efficiency of foreign gene expression. The product of the VA_I gene is a small RNA molecule which specifically inactivates a host cell enzyme known as dsRNA-activated inhibitory protein kinase (DAI kinase). This enzyme is induced by the presence of double-stranded RNA (dsRNA), which is sometimes generated adventitiously by dual transcription from plasmid DNA vectors. The enzyme blocks protein synthesis in the cell by phosphorylating and inactivating the host translational initiation factor eIF2. The presence of the VA_I RNA interrupts this process and therefore

Table 2.2   A selection of cationic polymer transfection reagents developed for gene transfer

Reagent or reagent family	Composition, if known	Supplier
CellFectin	Cationic lipopolyamine	Invitrogen
CytoPure	Cationic polymer, biodegradable	Qbiogene
ExGen family	Polyethyleneimine	MBI Fermentas
GeneJammer	Polyamine	Strategene
GeneJuice	Polyamine	Novagen
jetPEI	Polyethyleneimine	Qbiogene
Polyethylenimine-Transferrinfection	Polyethyleneimine	Bender MedSystems
PolyFect	Dendrimer	Qiagen
SuperFect	Dendrimer	Qiagen
Targetfect F2 and F4	Cationic polymer	Targeting Systems
TransIT family	Polyamine	Pan Vera Corp

facilitates protein synthesis at normal levels. Vectors carrying the adenoviral VA genes are commercially available, e.g. pAdVAntage, which is marketed by Promega. The adenoviral VA genes enhance expression independently of the transcriptional control elements used in the vector, since they act at the level of protein synthesis. Other reagents act at the level of transcription and only work with certain promoters and enhancers. For example, sodium butyrate is used to increase transient expression levels in transfection experiments with plasmids containing the SV40 early promoter and enhancer, since it enhances transcription from this regulator.

As stated earlier, chloroquine increases transfection efficiencies by inhibiting the acidification of endosomal vesicles, thus protecting DNA from degradation and promoting escape into the cytoplasm. More recently, various synthetic peptides have been developed which can enhance transfection efficiencies by disrupting endosomes, particularly when used in combination with lipid-based transfection methods. Examples include both natural peptides, e.g. the N-terminal sequence from hemagglutinin-A2 virus (GLFEAIAGFIENGWEGMIDGGGC) and synthetic peptides such as KALA (WEAKLAKALAKALAKHLAKALAKALKACEA). As well as such *endosomolytic peptides*, other peptides are fusogenic, i.e. they facilitate membrane penetration. Examples include peptides derived from HIV proteins gp41 (GALFLGFLGAAGST MGA) and Tat (YGRKKRRQRRR). Finally, some peptides help target the exogenous DNA to the nucleus, e.g. SV40 T antigen (PKKKRKV) and c-Myc (PAAKRVKL). (See references (11,12,28,29).)

Targeted transfection

A recent development in transfection technology, suitable for gene therapy applications, is the delivery of DNA to particular cells by conjugation to a specific ligand. The ligand interacts with receptors on the cell surface allowing both it and the attached DNA to be internalized. This strategy was first used to deliver plasmid DNA to liver cells by targeting the liver-specific asialoglycoprotein receptor. Plasmid DNA was conjugated to the ligand, asialoorosomucoid protein, using a polylysine bridge. The conjugate was added to cultured cells, and plasmid reporter gene activity was observed in liver cells but not other cell types. Since then, many similar experiments have been performed, and the efficient receptor-mediated transfection of many cell types using various ligands has been reported. A greater than 1000-fold enhancement of gene expression occurs if the ligand–DNA complexes are joined to adenoviral particles, which are known to disrupt endosomes as part of their infection strategy. More recently, adenovirus-derived peptides have been used for the same purpose, because these are less toxic and are less likely to provoke an immune response after in vivo delivery. Receptor-mediated transfection is highly efficient in cell culture, resulting in the transfection of up to 90% of cells carrying the appropriate receptor. Less success has been observed for in vivo gene transfer, partly because the ligand–nucleic acid complexes are degraded in serum, and partly because the size of the particles appears to be a critical parameter for the transfection of different cell types.

Another potential solution to the problem of targeted gene delivery is to control the physical and chemical properties of the gene delivery vehicle itself. This reflects the fact that some polymers change their properties in response to a physical stimulus. For example, a synthetic polymer has been described whose affinity to DNA changes on exposure to UV light. The polymer is based on azobenzene, and the azo moiety undergoes *trans*-II-*cis* isomerism following UV irradiation causing DNA to be released from the complex. In cultured cells, exposure to UV light increases transfection efficiency by up to 50%. Polymeric complexes responsive to UV light have limited use for gene therapy because it would be difficult to illuminate internal organs. However, another synthetic complex has been described that is responsive to heat. This polymer is based on N-isopropylacrylamide (PIPAAm) and undergoes a phase transition at a lower critical solution temperature of 32°C. Below 32°C, the polymer is hydrophilic and soluble, and forms a loose complex with DNA. Above 32°C, the polymer is hydrophobic and compact, and forms a tight complex by aggregation. A tight complex is preferable for DNA delivery because it is suitable for uptake and resistant to nucleases, whereas a loose complex is better for transcription because it facilitates access by transcription factors. Therefore, at normal body temperatures the DNA complex may be taken up efficiently by all cells, but poorly expressed. Local cooling, e.g. through the application of ice to the body surface or the use of catheters, can then induce gene expression in specific tissues or organs. The properties of the polymer can be modified by increasing the proportion of hydrophobic or hydrophilic chemical groups, thus lowering or raising the transition temperature, respectively. (See reference (30).)

2.3 Physical transfection methods

General principles

The major distinction between physical and chemical transfection is that in physical methods the DNA is delivered directly into either the cytoplasm or the nucleus using some kind of physical force, without any requirement for interaction with the plasma membrane. This avoids involvement with the endosomal pathway and thus limits the amount of damage sustained by the exogenous DNA. Other than that, there is really very little in common between the various physical transfection methods, because the physical principles and delivery routes vary considerably from case to case. One further uniting factor is that all the physical transfection methods employ some sort of apparatus, usually expensive, which is required to administer the physical force. The DNA may be introduced free in solution, but it may be beneficial to protect it by forming chemical complexes (e.g. with polyamines) to reduce the damaging effects of shear forces during the transfer process. (See references (21,22,31).)

Electroporation

Electroporation is the transfection of cells following their exposure to a pulsed electric field. This causes a number of nanometer-sized pores to open in the plasma membrane for up to 30 minutes, allowing the uptake of free DNA from the surrounding medium. Afterwards, the pores close spontaneously with no noticeable adverse effects on the treated cells. This method, first used with animal cells by Neumann and colleagues in 1982 (32), is ideal for many established cell lines, especially those recalcitrant to chemical transfection methods. It is efficient, highly reproducible, suitable for both stable and transient transfection, and has the added advantage that transgene copy number can be at least partially controlled. Electroporation was initially only applicable to cells that could be cultured in suspension, but an adaptation of the technique allowing it to be applied to cells on polyethylene terephthalate or polyester membranes has ensured that it can also be used with cells growing in monolayers. Most recently, a modification of the technique known as nucleofection has been described and developed as the Nucleofector platform by the German biotechnology company amaxa GmbH. This method combines specific electroporation conditions with different transfection reagents to promote direct electroporation-mediated gene transfer to the nucleus, resulting in the efficient transfection of traditionally difficult targets including primary cells.

The standard electroporation procedure is very simple (*Protocol 2.4*). Cells are suspended in or flooded with an electroporation buffer and exposed to a brief, high voltage electrical pulse. The magnitude and duration of the pulse determine the transfection efficiency, and these conditions must be established empirically for different cell lines. Many cells are efficiently electroporated using a brief high-voltage pulse (800–1500 V), but others, especially primary cells, may be killed by such treatments and respond better to a longer-lasting pulse of 100–300 V. Disadvantages of this technique include the requirement for specialized capacitor discharge equipment capable of accurately controlling pulse length and voltage (e.g. BioRad Gene Pulser and Gene Pulser II), the requirement for larger numbers of cells and higher DNA concentrations than used in chemical transfection methods, and the rather high level of cell death that accompanies this procedure.

Electroporation has also been explored as a method for in vivo gene transfer. It has been used with success to introduce DNA into surface or near-surface tissues such as skin, muscle and melanoma, and also into internal organs such as the liver. In each case, DNA was injected into the target tissue using a conventional needle and electroporation was achieved through the use of needle electrodes. However, in at least one case electroporation has been carried out by the direct application of electrodes to the skin following shaving and mild abrasion. Among the commercial devices available for electroporation, at least one has been designed specifically for electroporation-mediated gene therapy in humans. (See references (32–34).)

Laser poration

Another in vitro transfection technology based on pore formation, which has not become widely used, is transfection following laser treatment. This involves a similar DNA uptake mechanism to electroporation, i.e. free DNA is taken up directly from the surrounding medium through transient pores created by a finely focused laser beam. Like microinjection (see below) this strategy can be applied only to a small number of cells at a time, but with optimal DNA concentration can result in stable transfection frequencies of greater than 0.5%. (See reference (40).)

Microinjection

The direct microinjection of DNA into the cytoplasm or nuclei of cultured cells is sometimes used as a transfection method. Although highly efficient at the level of individual cells, this procedure is time consuming and only a small number of cells can be treated. Originally, this technique was used for the transformation of cells that were resistant to any other method of transfection. Stable transfection efficiencies are extremely high, in the order of 20%, and very small quantities of DNA are sufficient. Another major advantage of this technique is that direct nuclear delivery is possible, avoiding the inefficient endogenous pathway for transporting DNA to the nucleus, and also ensuring that the DNA is delivered intact. Although many reports describe the use of microinjection in cultured cells, the most significant use of this technique remains the introduction of DNA into the oocytes, eggs and embryos of animals, either for transient expression analysis (e.g. in fish or *Xenopus*) or to generate transgenic animals (e.g. mice, *Drosophila*). Microinjection is suitable for the introduction of large vectors such as YACs into the pronuclei of fertilized mouse eggs, but since DNA delivered in this manner must be very pure, painstaking preparation is necessary to avoid fragmentation. Shearing can also occur in the delivery needle, and large DNA fragments are often protected by suspension in a high salt buffer and/or mixing with polyamines and other protective agents. An important improvement in this technique for the transfection of cultured cells was automation, with computer-controlled micromanipulation and microinjection processes as well as the automated production of injection capillaries and the standardization of cell preparation procedures. The development of computer-assisted and microprocessor-controlled injection systems makes high injection rates feasible and allows for quantitative microinjection with optimal reproducibility. (See references (41–43).)

Transfection by particle bombardment

Particle bombardment (also known as biolistics or microprojectile transfection) is a relatively recent addition to the range of transfection techniques available to scientists working with animal cells. The procedure involves coating micrometer-sized gold or tungsten particles with DNA and then accelerating the particles into cells or tissues. A major advantage of this method is that DNA can be delivered to deep cells in tissue slices, and the depth of penetration can be adjusted by changing the applied force. The size and total mass of the particles and the force of the bombardment are important parameters that balance efficient penetration against cell damage.

The technique was developed for the transformation of maize and is now a method of choice for generating transgenic cereal plants. For animal cells, the technique has been less widely used because it is usually simpler to transfect cultured cells by alternative well-established methods. However, the technique has found a role in the transfection of whole organs and tissue slices, and more recently for the transfer of DNA to surface organs in gene therapy. Particle bombardment is much less efficient than traditional injection for in vivo muscle transfection, but has nevertheless allowed the robust expression of several viral and bacterial antigens, resulting in a sustained immune response.

At the current time, the only commercially available bombardment devices are those produced by BioRad, which are based on the use of high-pressure helium as the productive force. The PST1000/He bench-mounted device comprises a vacuum chamber into which the target material must be inserted prior to transfection (*Figure 2.5*) while the hand-held Helios gun can be used without a vacuum (*Figure 2.6*). Although the former has been successful for the transfection of cells in tissue slices, it is limited to those tissue samples that will fit in the relatively small vacuum chamber and are not damaged in a vacuum. The hand-held gun is more versatile, but in both cases the efficiency of transfection comes at the cost of increased tissue damage. The most recent developments of the technique have focused on instrument modifications that allow gas pressures to be reduced from 200 psi to approximately 50 dpi, resulting in less tissue damage. A variety of other proprietary devices have also been developed, including the Accell gun which is driven by electrical discharge, pneumatic guns and guns based

Figure 2.5

The biolistic PDS-1000/He unit. Top, main chamber containing the microcarrier launch assembly and the bombardment helium pressure gauge. The central gauge (in the left side of the instrument) monitors the vacuum within the chamber, and the two lower knobs adjust the vacuum flow and vent rates. The helium-metering valve is next to the main chamber (lower right). Permission to use the materials has been granted by Bio-Rad Laboratories, Inc.

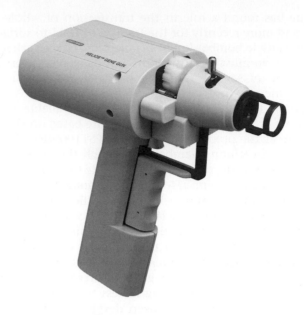

Figure 2.6

The Helios gene gun is a convenient, hand-held device that provides rapid and direct gene transfer into a range of targets in vivo. The unit uses an adjustable low-pressure helium pulse to sweep DNA- or RNA-coated gold microcarriers from the inner wall of a small plastic cartridge directly into the target. Permission to use the materials has been granted by Bio-Rad Laboratories, Inc.

on flowing helium, but few of these are routinely applied to the transfection of animal cells. For in vivo gene transfer, a proprietary device for gene transfer to skin cells has been developed known as the needle-free PowderJect™ system. A DNA vaccine against the hepatitis B virus delivered into skin cells using this system has demonstrated the induction of both humoral and cell-mediated immune responses in humans. At this point, experience with particle bombardment gene therapy is still limited because the gene guns made by Bio-Rad are for research use only. (See references (44–46).)

Transfection by ultrasound

Ultrasound transfection involves the exposure of cells to a rapidly oscillating probe, such as the tip of a sonicator. The transfection mechanism is similar in some ways to electroporation in that the application of ultrasound waves to a dish of cells or a particular tissue results in the formation and collapse of bubbles in the liquid, including the cell membrane, a process known as cavitation. The transient appearance of such cavities allows DNA to cross the membrane into the cytoplasm. It has been shown that the application of low-frequency ultrasound allows the efficient delivery of nucleic acids into mammalian cells both in vitro and in vivo, because the plasmid DNA is left structurally intact. Furthermore, ultrasound-mediated gene delivery raises no safety concerns because the ultrasound waves appear to have no adverse effects when focused on different anatomic locations in the human body. Gene transfer in vivo is generally achieved by injection followed by the application of a focused ultrasound device. (See references (47,48).)

Injection of naked DNA

Naked DNA transfects cultured cells very inefficiently when added to the growth medium, due both to degradation in the culture medium and poor internalization. Similarly, uncomplexed DNA administered intravenously to live animals is rapidly degraded following hepatic uptake. Unlikely as it may seem, however, the injection of uncomplexed plasmid DNA in vivo into muscle and skin results in high-efficiency gene transfer and expression[3]. This can be exploited to raise a protective immune response against viral and other pathogenic antigens, and against tumor proteins. The application of naked DNA to other tissues has also resulted in successful transfection, e.g. using balloon catheters to coat the internal walls of arteries. DNA has also been injected into the blastocoel cavity of mouse embryos, resulting in the generation of transgenic mice, albeit with much lower efficiency than pronuclear microinjection (Chapter 5). It has even proven possible to inject DNA directly into tumors and derive a therapeutic effect. The mechanism by which naked DNA is taken into, e.g. muscle cells is entirely unknown. Perhaps surprisingly, dry DNA pellets are more efficient transfection substrates than aqueous solutions, and the pre-treatment of muscles with hypertonic solutions can also increase the efficiency of DNA transfer (perhaps by an analogous process to the glycerol shock used to increase transfection efficiencies in cultured cells). It is thought that the extracellular DNA may become coated with particular extracellular proteins, which form a positively charged complex and promote interaction with cell membranes and uptake by receptor-mediated endocytosis.

One variation on conventional injection is the jet delivery system. The jet gun forces a DNA solution through a small orifice that creates a very fine, high-pressure stream to penetrate the skin, depositing naked DNA in the tissue beneath. Jet injection of DNA into muscle and skin has been shown to facilitate DNA uptake and high-level expression. Another variation is the hydrodynamic gene delivery system, which involves the rapid injection of very large volumes of naked DNA directly into the bloodstream to ensure that at least some survives degradation by nucleases. This has been shown to facilitate gene transfer to a variety of internal organs in rodents and primates, with the highest expression levels in the liver. (See references (49,50).)

---

3 Note that the delivery of DNA by conventional needle injection is only physical delivery to the *tissue*. There is no evidence that the needle facilitates gene transfer into any of the cells at the injection site. What appears to happen is that the local *extracellular* concentration of DNA becomes very high and this stimulates DNA uptake in the manner of a conventional chemical transfection technique.

2.4 Plasmid vector systems for transient transfection

Transient transfection with non-replicating vectors

In transient transfection, DNA is introduced into the cell but is not expected to integrate into the genome. Instead, the DNA remains in the nucleus for a short period of time in an extrachromosomal state. Almost any vector and any cell line can be used for this type of transient transfection, and recombinant proteins can be synthesized as long as the vector contains a functional expression cassette. The onset of protein expression is rapid, but the yield depends not only on the regulatory elements in the vector, but also on the transfection efficiency (i.e. how much DNA is taken into each cell) and the stability of the extrachromosomal DNA (i.e. how long it lasts once inside). Covalently closed circular plasmid DNA survives 12–48 hours in many cell lines before it is degraded and diluted, allowing the transient expression of foreign genes from simple vectors lacking eukaryotic maintenance sequences. Some cell lines, however, are renowned for the stability of transfected DNA, allowing survival and foreign gene expression for over 80 hours. One example is the adenovirus-transformed human embryonic kidney (HEK) line 293. Covalently closed and super-coiled plasmid DNA is required for efficient transient transfection. Relaxed DNA (i.e. nicked circles or linear DNA) is a poor template for transcription, and is prone to exonucleolytic degradation. Ideally, high-quality plasmid DNA, prepared using either cesium chloride equilibrium density gradient ultracentrifugation or anion exchange chromatography, should be used for transient transfection experiments. (See references (2,51).)

Transient transfection with polyomavirus replicons

The polyomaviruses are small DNA viruses (family Papovoviridae) which have non-enveloped, icosahedral capsids and circular double-stranded DNA genomes approximately 5 kbp in size (*Figure 2.7*). The best-characterized species is simian virus 40 (SV40), whose productive host range is limited to certain monkey cells. SV40 was an early model system studied in great detail by molecular biologists in the 1970s, and was the first animal virus to be exploited as a vector. The initial SV40-based vectors were viral transduction vectors, but these are no longer used because of their very limited capacity for foreign DNA (maximum insert size 2.5 kbp), reflecting the small size of the capsid.

A breakthrough came with the discovery that plasmids containing a polyomavirus origin of replication could be propagated in the same way as the virus genome, i.e. as episomal replicons in permissive cells. The replication of polyomavirus-derived replicons requires the presence of a viral-encoded regulator termed the T-antigen, but in the absence of a viral capsid there is no effective limit to insert size. A number of polyomaviruses have been exploited to develop episomal vectors, including SV40 for monkey cells, murine polyomavirus for mouse cells, and the human BK polyomavirus for human cells. The related bovine papillomavirus (BPV) has also been used in murine cells. BK- and BPV-derived replicons can be maintained episomally at a moderate copy number and allow long-term protein expression

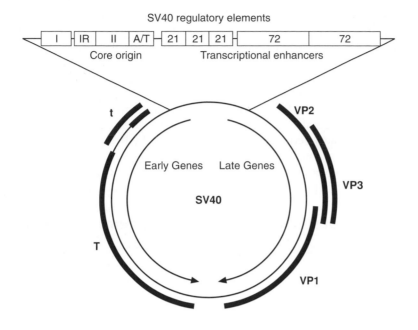

SV40 regulatory elements

| I | IR | II | A/T | 21 | 21 | 21 | 72 | 72 |

Core origin   Transcriptional enhancers

Figure 2.7

Map of the SV0 genome.

(see below). Conversely, the replication of SV40- and murine polyomavirus-derived vectors is uncontrolled. In permissive cells, with readily available T-antigen, such vectors can produce up to 10^5 vector molecules per cell in a matter of hours. This facilitates high-level recombinant protein synthesis, making such host–vector systems among the most efficient available. Unfortunately, the high level of replication and gene expression cannot be sustained by the host cell, and cell death occurs within a few days. SV40 and polyomavirus replicons are therefore suitable only for transient transfection, not because the DNA is lost or degraded, but because so much foreign DNA is synthesized that the cell cannot survive.

In some polyomavirus vectors, the T-antigen coding sequence is included within the vector, enabling propagation in any permissive cell line. Other vectors lack the T-antigen, which must then be supplied from another source. The development of the COS cell line simplified the use of SV40-based vectors because this monkey cell line is stably transformed with a partial SV40 genome. The name means cell line <u>C</u>V-1 (the African Green Monkey cell line from which COS cells are derived), <u>o</u>rigin of replication and <u>S</u>V40. The integrated SV40 origin is mutated and non-functional, but the early transcription unit is functional allowing constitutive expression of the T-antigen. This can stimulate the replication of any SV40 replicon in the same nucleus, and therefore allows the episomal propagation of SV40 vectors lacking their own T-antigen transcription unit. A whole series of SV40-based expression vectors has been designed for use in COS cells. Cell lines have also been developed for the propagation of murine polyomavirus replicons, e.g. the mouse cell line MOP-8. Notably, while the SV40 origin of replication may function only in certain permissive monkey cells, the promoter and enhancer sequences of this and other viruses are extremely promiscuous. SV40 replicons can therefore be used for transient transfection of any mammalian cell type although protein yields are much lower than for monkey cells because there is no episomal replication. In non-permissive cells, SV40 DNA may also integrate into the genome resulting in stable transformation. More recently, it has been possible to generate permissive monkey cell lines stably transformed with episomal SV40-based vectors by using conditionally expressed or temperature sensitive T-antigens, which sponsor more moderate and tolerable replication rates. (See references (52,53).)

2.5   Episomal stable transformation systems

Replicating, extrachromosomal vectors for stable transformation possess a number of distinct advantages over integrating vectors, including the absence of the position and dosage effects often seen in integrated trans-genes (Chapter 8), and the high stable transformation efficiency. The latter reflects the fact that stable transfection by episomal maintenance is equiv-alent in efficiency to transient transfection with any other vector (up to 10^5 times more efficient than stable transfection by integration) if a selectable marker is included in the vector to facilitate maintenance following trans-fection. Episomal vectors can also be separated easily from chromosomal DNA, which is particularly useful in techniques such as expression cloning that rely on the recovery of cloned DNA. The genomes of three viruses have been widely exploited to construct episomal plasmid vectors: BK virus, bovine papillomavirus (BPV) and Epstein–Barr virus (EBV). (See references (54–58).)

BK virus replicons

The human BK polyomavirus is stably maintained at a moderate copy number in human cells (approximately 500 copies per cell), and plasmid vectors carrying the BK virus origin of replication are similarly maintained as long as the BK virus T-antigen is available. BK virus-derived plasmid vectors contain BK replicon functions, a selectable marker (e.g. *neo*) and a mammalian transcription unit. This allows high-level recombinant protein expression to be achieved by increasing the concentration of selective agent (e.g. G418) in the medium and hence selecting for increased replicon copy number. In this way, stable cell lines carrying up to 9000 copies of the vector can be generated and propagated for years, although the replicon copy number drops when selection is withdrawn.

BPV replicons

Bovine papillomavirus (BPV) productively infects cattle and certain other domestic mammals. It can also infect cultured mouse cells, but cannot produce progeny virions. Instead, 50–100 copies of the genome are main-tained episomally as plasmids, and it is this property which has been exploited to develop a series of episomal expression vectors. The earliest BPV-derived vectors comprised the entire BPV genome cloned in a bacterial plasmid with a mammalian expression cassette. Since the wild-type BPV genome causes oncogenic transformation, the transfected cells could be identified by their proliferative behavior. The early functions of the virus are carried on a 5.5 kbp sector of the BPV genome, which is termed the 69% transforming fragment (BPV$_{69T}$), and this appears to be sufficient for establishment and maintenance of the episomal state. This fragment forms the basis of a more versatile series of expression vectors that also contain a mammalian selectable marker to extend the range of possible host cells. The ability of BPV$_{69T}$ vectors to replicate efficiently is increased in some but not all cells by certain mammalian genomic sequences, which are included, e.g. in the BV-1 vector series. BPV vectors can establish long-term, moderate-level foreign gene expression in a broad range of mammalian cell types, although 3T3 and C127 fibroblasts are the most popular hosts. There is no limit to insert size, allowing both cDNAs and full genes to be expressed. However, BPV vectors suffer from several limitations including their large size compared to standard plasmids and the consequent loss of cloning versatility, and their low stability, which means that recombinant vectors may delete their cargo of foreign DNA. Recombination between vectors can also occur, so that either single plasmids or oligomers may be propagated. These effects are largely unpredictable, depending on multiple parameters such as cell line, vector type and insert structure. The same parameters also affect replication efficiency, so the episomal copy number may vary from very low (20 plasmids per cell) to over 300 copies.

Epstein–Barr virus replicons

Due to the instability of BPV replicons, many researchers have turned to Epstein–Barr virus (EBV) as an episomal expression vector. EBV is a human herpes virus whose productive host range is limited to primates and a few other mammals. The virus is naturally lymphotropic and commonly infects B-cells, causing infectious mononucleosis. The virus has a large, double-stranded linear DNA genome that circularizes shortly after penetration by interaction between terminal repeat sequences. In culture, the virus can cause lytic infection, resulting in cell death, but most infected cells are transformed into a proliferative state and the virus genome is maintained as a latent, episomally replicating plasmid, with a copy number under 1000.

Although the host range of the virus is limited by its interaction with a specific receptor found on very few cell types, the circular EBV genome is maintained in many primate cells following transfection, and thus shows great potential as a broad cellular host-range expression vector. However, the species host range is still limited to primate and some canine cells – rodent cell lines are not permissive for episomal EBV replication. Only two regions of the EBV genome are essential for the establishment and maintenance of latent replication: the bipartite origin of replication *oriP*, and the gene encoding the *trans*-acting regulator of transcription and replication, EBV nuclear antigen-1 (*EBNA-1*). These sequences form the basis of EBV latent expression vectors, which are maintained at a copy number of 2–50 per cell. A distinct origin, *ori*$_{Lyt}$, and a different regulator termed ZEBRA, are required for lytic replication, thus *oriP*/EBNA-1 vectors do not increase in copy number in cells undergoing lytic EBV infection. More recently developed EBV vectors containing both origins can be maintained as low copy number vectors in latently infected B-cell lines, and amplified approximately 500-fold when cells are transfected with constitutively expressed *BZLF-1* (the ZEBRA gene) to induce lytic replication. EBV vectors may contain *oriP* only (in which case EBNA-1 must be supplied in *trans*) or both *oriP* and the EBNA-1 coding region driven by a mammalian promoter (in which case the vectors are helper independent). EBV vectors must also carry a selectable marker because the transformation efficiency can be lower than 10%, and in the absence of selection, both *oriP* and *oriP*/EBNA-1 vectors are passively lost at the rate of 2–6% of cells per generation. EBV vectors have been developed for expression cloning, since they can be recovered as intact elements from mammalian cells, and can be maintained in *E. coli* as cosmids if bacteriophage λ *cos* sites are included. EBV replicons also form the basis of one form of mammalian artificial chromosome, the human artificial episomal chromosome.

References

1. Liu D, Chiao EF, Tian H (2004) Chemical methods for DNA delivery: An overview. In: Heiser WC (ed.) (2004) *Gene Delivery to Mammalian Cells. Volume 1 – Nonviral Gene Transfer Techniques*, pp 3–23. Humana Press, Towata, NJ.
2. Keown WA, Campbell CR, Kucherlapati RS (1990) Methods for introducing DNA into mammalian cells. *Methods Enzymol* **185**: 527–537.
3. Graham FL, Van der Eb AJ (1973) A new technique for the assay of infectivity of human adenovirus 5 DNA. *Virology* 52: 456–460.
4. Szybalska E, Szybalski W (1962) Genetics of human cell line. IV. DNA-mediated hereditable transformation of a biochemical trait. *Proc Natl Acad Sci USA* **48**: 2026–2034.
5. Wigler M, Silverstein S, Lee LS, Pellicer A, Cheng Y, Axel R (1977) Transfer of purified herpes virus thymidine kinase gene to cultured mouse cells. *Cell* **11**: 223–232.
6. Chen C, Okayama H (1987) High efficiency transformation of mammalian cells by plasmid DNA. *Mol Cell Biol* 7: 2745–2751.
7. Chen C, Okayama H (1988) Calcium phosphate-mediated gene transfer: a highly efficient transfection system for stably transforming cells with plasmid DNA. *Biotechniques* **6**: 632.
8. Jordan M, Schallhorn A, Wurm FW (1996) Transfecting mammalian cells: Optimization of critical parameters affecting calcium-phosphate precipitate formation. *Nucleic Acids Res* **24**: 596–601.
9. Jordan M, Wurm F (2004) Transfection of adherent and suspended cells by calcium phosphate. *Methods* 33: 136–143.
10. Pagano JS, Vaheri A (1965) Enhancement of infectivity of poliovirus RNA with diethylaminoethyl-dextran (DEAE-D). *Arch Gesamte Virusforsch* **17**: 456–464.
11. Lopata MA, Cleveland DW, Sollner-Webb B (1984) High-Level Transient Expression of a chloramphenicol acetyl transferase gene by deae-dextran mediated DNA transfection coupled with a dimethylsulfoxide or glycerol shock-treatment. *Nucleic Acids Res* 12: 5707–5717.
12. Sussman DJ, Milman G (1984) Short-term, High-efficiency expression of transfected DNA. *Mol Cell Biol* 4: 1641–1643.
13. Schaefer-Ridder M, Wang Y, Hofschneider PH (1982) Liposomes as gene carriers – Efficient transformation of mouse L-cells by thymidine kinase gene. *Science* **215**: 166–168.

14. Tseng W-C, Huang L (1998) Liposome-based gene therapy. *Pharmaceut Sci Technol Today* **1**: 206–213.
15. Felgner PL, Gadek TR, Holm M, Roman R, Chan HW, Wenz M, Northrop JP, Ringold GM, Danielson M (1987) Lipofection: a highly efficient, lipid-mediated DNA transfection procedure. *Proc Natl Acad Sci USA* **84**: 7413–7417.
16. Byk G, Dubertret C, Escriou V, *et al.* (1998) Synthesis, activity, and structure-activity relationship studies of novel cationic lipids for DNA transfer. *J Med Chem* **41**: 224–235.
17. Hui SW, Langner M, Zhao YL, Ross P, Hurley E, Chan K (1996) The role of helper lipids in cationic liposome-mediated gene transfer. *Biophys J* **71**: 590–599.
18. Templeton NS, Lasic DD, Frederik PM, Strey HH, Roberts DD, Pavlakis GN (1997) Improved DNA: liposome complexes for increased systemic delivery and gene expression. *Nat Biotechnol* **15**: 647–652.
19. Rose PC, Hui SW (1999) Lipoplex size is the major determinant of in vitro lipofection efficiency. *Gene Ther* **6**: 651–659.
20. Miller AD (1998) Cationic liposomes for gene therapy. *Angew Chem Int Ed Engl* **37**: 1768–1785.
21. Li S, Ma Z (2001) Non-viral gene therapy. *Curr Gene Therapy* **1**: 201–226.
22. Niidome T, Huang L (2002) Gene therapy progress and prospects: Nonviral vectors. *Gene Ther* **9**: 1647–1652.
23. Davis ME (2002) Non-viral gene delivery systems. *Curr Opin Biotechnol* **13**: 128–131.
24. Boussif O, Lezoualch F, Zanta MA, Mergny MD, Scherman D, Demeneix B, Behr JP (1995) A versatile vector for gene and oligonucleotide transfer into cells in culture and in vivo: polyethylenimine. *Proc Natl Acad Sci USA* **92**: 7297–7301.
25. Tang MX, Redemann CT, Szoka FC Jr (1996) In vitro gene delivery by degraded polyaminediamine dendrimers. *Bioconj Chem* **7**: 703–714.
26. Kukowska-Latallo JF, Bielinska AU, Johnson J, Spindler R, Tomalia DA, Baker JR Jr (1996) Efficient transfer of genetic material into mammalian cells using Starburst polyamidoamine dendrimers. *Proc Natl Acad Sci USA* **93**: 4897–4902.
27. Tang MX, Szoka FC (1997) The influence of polymer structure on the interactions of cationic polymers with DNA and morphology of the resulting complexes. *Gene Ther* **4**: 823–832.
28. Akusjarvi G, Svensson C, Nygard O (1987) A mechanism by which adenovirus virus-associated RNA$_I$ controls translation in a transient expression assay. *Mol Cell Biol.* **7**: 549–551.
29. Zhang L, Ambulos N, Mixon AJ (2004) DNA delivery to cells in culture using peptides In: Heiser WC (ed.) *Gene Delivery to Mammalian Cells. Volume 1 – Nonviral Gene Transfer Techniques*, pp 33–52. Humana Press, Towata, NJ.
30. Yokoyama M (2002) Gene delivery using temperature-responsive polymeric carriers. *Drug Discovery Today* **7**: 426–432.
31. Chou TW, Biswas S, Lu S (2004) Gene delivery using physical methods: An overview. In: Heiser WC (ed.) *Gene Delivery to Mammalian Cells. Volume 1 – Nonviral Gene Transfer Techniques*, pp 147–165. Humana Press, Towata, NJ.
32. Neumann E, Schaefer-Ridder M, Wang Y, Hofschneider PH (1982) Gene transfer into mouse lyoma cells by electroporation in high electric fields. *EMBO J* **1**: 841–845.
33. Yang TA, Heiser WC, Sedivy JM (1995) Efficient in situ electroporation of mammalian cells grown on microporous membranes. *Nucleic Acids Res* **23**: 2803–2810.
34. Gresch O, Engel FB, Nesic D, *et al.* (2004) New non-viral method for gene transfer to primary cells. *Methods* **33**: 151–163.
35. Chu G, Hayakawa H, Berg P (1987) Electroporation for the efficient transfection of mammalian cells with DNA. *Nucleic Acids Res* **15**: 1311–1326.

36. Knutson JC, Yee D (1987) Electroporation: parameters affecting transfer of DNA into mammalian cells. *Anal Biochem* **164**: 44–52.

37. Titomirov AV, Sukharev S, Kistanova E (1991) In vivo electroporation and stable transformation of skin cells of newborn mice by plasmid DNA. *Biochem Biophys Acta* **1088**: 131–134.

38. Aihara H, Miyazaki J (1998) Gene transfer into muscle by electroporation in vivo. *Nat Biotechnol* **16**: 867–870.

39. Bigey P, Bureau MF, Scherman D (2002) In vivo plasmid DNA electrotransfer. *Curr Opin Biotechnol* **13**: 443–447.

40. Kurata S, Tsukakoshi M, Kasuya T, Ikawa Y (1986) The laser method for efficient introduction of foreign DNA into cultured-cells. *Exp. Cell Res.* **162**: 372–378.

41. Graessmann M, Graessmann A (1983) Microinjection of tissue culture cells. *Methods Enzymol* **101**: 482–492.

42. Ansorge W (1982) Improved system for capillary microinjection into living cells. *Exp Cell Res* **140**: 31–37.

43. Ansorge W, Pepperkok R (1988) Performance of an automated system for capillary microinjection into living cells. *J Biochem Biophys Methods* **16**: 283–292.

44. Burkholder JK, Decker J, Yang NS (1993) Rapid transgene expression in lymphocyte and macrophage primary cultures after particle bombardment-mediated gene transfer. *J Immunol Methods* **165**: 149–156.

45. Lo DC, McAllister AK, Katz LC (1994) Neuronal transfection in brain slices using particle-mediated gene transfer. *Neuron* **13**: 1263–1268.

46. O'Brien J, Lummis SCR (2004) Biolistic and diolistic transfection: using the gene gun to deliver DNA and lipophilic dyes into mammalian cells. *Methods* **33**: 121–125.

47. Mitragotri S, Blankschtein D, Langer R (1996) Transdermal drug delivery using low-frequency sonophoresis. *Pharm Res* **13**: 411–420.

48. Wyber JA, Andrews J, D'Emanuele A (1997) The use of sonication for the efficient delivery of plasmid DNA into cells. *Pharm Res* **14**: 750–756.

49. Wolff JA, Malone R, Williams WP, Chong W, Acsadi G, Jani A, Felgner PL (1990) Direct gene transfer into mouse muscle in vivo. *Science* **247**: 1465–1468.

50. Furth PA, Shamay A, Wall RJ, Hennighausen L (1992) Gene transfer into somatic tissues by jet injection. *Anal Biochem* **205**: 365–368.

51. Chisholm V (1995) High efficiency gene transfer to mammalian cells. In: Glover DM, Hames BD (eds) *DNA Cloning: A Practical Approach. Volume 4, Mammalian Systems* (second edition), pp 1–42. IRL Press, Oxford, UK.

52. Kaufman RJ (1990) Vectors used for expression in mammalian cells. *Methods Enzymol* **185**: 487–511.

53. Aruffo A (1991) Transient expression of proteins using COS cells. In: Ausubel RBFM, Kingston RE, Moore DD, Seidman JG, Smith JA, Struhl K (eds) *Current Protocols in Molecular Biology*, pp. 16.13.1–16.13.7. New York. John Wiley and Sons Inc.

54. Van Craenenbroeck K, Vanhoenacker P, Haegeman G (2000) Episomal vectors for gene expression in mammalian cells. *Eur J Biochem* **267**: 5665–5668.

55. Milanesi G, Barbanti-Brodano G, Negrini M, Lee D, Corallini A, Caputo A, Grossi MP, Ricciardi RP (1984) BK virus plasmid expression vector that persists episomally in human cells and shuttles into *Escherichia coli*. *Mol Cell Biol* **4**: 1551–1560.

56. DiMaio D (1987) Papillomavirus cloning vectors. In: Salzman NP, Howley PM (eds) *The Papoviridae: The Papillomaviruses*, volume 2. Plenum Publishing, NY.

57. Margolskee RF (1992) Epstein-Barr virus-based expression vectors. *Curr Top Microbiol Immunol* **185**: 67–95.

58. Sun TQ, Fernstermacher DA, Vos JM (1994) Human artificial episomal chromosomes for cloning large DNA fragments in human cells. *Nature Genet* **8**: 33–41.

Protocols

Contents

MATERIALS

General	High-quality plasmid DNA at 1 mg ml^{-1}
	3 M sodium acetate (pH 4.8)
	Ethanol (70%, absolute)
	Cells for transfection, exponentially growing
	NUNC or similar six-well culture plates
	Culture medium and any necessary additives
	Incubator
	Laminar flow hood
	Bench centrifuge, Sorvall H1000B or similar
Protocol 2.1	2.5 M CaCl$_2$
	2× HEPES-buffered saline (50 mM HEPES, 140 mM NaCl, 1.5 mM Na$_2$HPO$_4$; pH 7.05, filter sterilized)
Protocol 2.2	0.25 M CaCl$_2$
	2× BES-buffered saline (50 mM BES, 280 mM NaCl, 1.5 mM Na$_2$HPO$_4$; pH 6.96, filter sterilized)
Protocol 2.3	LipofectAMINE 2000 reagent (Invitrogen)
	Optimem 1 medium (Invitrogen)
Protocol 2.4	PBS or appropriate electroporation buffer from supplier
	Electroporator (e.g. Bio-Rad GenePulser, Eppendorf multiporator) and a supply of cuvettes

Protocol 2.1 Transfection of adherent cells using calcium phosphate (HEPES method)

1. Prepare as many six-well plates as necessary for the experiment and treat them as appropriate for the cell line being used – this may, for example, involve coating of the plates with poly-L-lysine or Matrigel. It is a good idea to label the wells underneath with the name of the DNA construct, to ensure the transfected cells can be identified after the experiment.

2. Plate approximately 2×10^5 cells in each well and incubate under the appropriate conditions for the cell line for 24 hours. This ensures that the cells are growing exponentially at the time of transfection.

3. Prepare DNA for transfection either by CsCl ultracentrifugation or using a proprietary column system, such as the QIAGEN maxi-prep kit. Precipitate the DNA in 0.1 volumes of 3 M sodium acetate (pH 4.8) and 2.5 volumes of absolute ethanol, mix briefly, pellet at 13 000 rpm using a standard bench centrifuge, wash with 70% ethanol and resuspend in sterile TE buffer or water in an Eppendorf tube. Determine the concentration by spectrophotometry and divide into 25-μg aliquots for transfection.

4. One hour before transfection, refresh the cells with 3 ml of medium per well. Transfer the DNA aliquots, $CaCl_2$ and 2× HEPES buffer to the flow hood.

5. Take cells out of the incubator and place under the flow hood. Prepare the calcium phosphate–DNA coprecipitate as follows: add 100 μl of 2.5 M $CaCl_2$ to each 25-μg aliquot of plasmid DNA in a sterile Eppendorf tube and bring a final volume of 1 ml with 0.1× TE (pH 7.6) or sterile water.

6. Add 150 μl of the calcium–DNA solution dropwise into an equal volume of 2× HEPES-buffered saline at room temperature, tapping the side of the tube to mix. Allow the solution to stand for 30 seconds to 1 minute. A visible precipitate will form.

7. Aspirate the coprecipitate and add, dropwise into the appropriate well. Rock the plate gently to mix the medium, which will become yellow-orange and turbid. Carry out this step as quickly as possible because the efficiency of transfection declines rapidly once the DNA precipitate is formed.

8. Return the plates to the incubator and leave for 2–4 hours, then remove the medium and refresh with 5 ml of prewarmed medium. Return to the incubator for 2–4 days.

9. For transient expression experiments, the cells can be assayed the following day. For stable transformation, replace the standard medium with selective medium and refresh daily for 1–2 weeks.

Protocol 2.2: Transfection of adherent cells using calcium phosphate (BES method)

This method is particularly suitable for the production of stable transformants using supercoiled plasmid DNA.

1. Prepare cells for transfection and DNA constructs as in Protocol 2.1.

2. Mix a 25-µg aliquot of plasmid DNA with 0.5 ml of 0.25 M $CaCl_2$ and add 0.5 ml of 2× BES-buffered saline (BBS). Mix gently and allow to stand for 20 minutes. A visible precipitate will not form.

3. Add 300 µl of the mixture dropwise to the cells, swirling gently. Return cells to the incubator for 24 hours.

4. Remove the medium and refresh with prewarmed medium. Incubate for a further 24 hours.

5. Trypsinize and replate the cells at the appropriate density, which should be determined empirically using reporter vectors. The density should be lower for cells that are transformed with a higher efficiency, and higher for those that are more difficult to transfect. Refresh with selective medium every 3 days.

Protocol 2.3: Transfection of mammalian cells using LipofectAMINE 2000

1. Prepare as many six-well plates as necessary for the experiment and treat them as appropriate for the cell line being used – this may, for example, involve coating of the plates with poly-L-lysine or Matrigel. It is a good idea to label the wells underneath with the name of the DNA construct, to ensure the transfected cells can be identified after the experiment.

2. Plate approximately 2×10^5 cells in each well and incubate under the appropriate conditions for the cell line for 48 hours.

3. Prepare DNA for transfection either by CsCl ultracentrifugation or using a proprietary column system, such as the QIAGEN maxi-prep kit. Precipitate the DNA in 0.1 volumes of 3 M sodium acetate (pH 4.8) and 2.5 volumes of absolute ethanol, mix briefly, pellet at 13 000 rpm using a standard bench centrifuge, wash with 70% ethanol and resuspend in sterile TE buffer or water in an Eppendorf tube. Determine the concentration by spectrophotometry and divide into 5-µg aliquots for transfection. Precipitate the DNA as described above and again wash in 70% ethanol. The pellet can be stored under 70% ethanol, can be air-dried, or it can be resuspended in 10 µl of PBS ready for transfection. Store at –20°C until required. Make sure the tubes are labeled carefully, especially if multiple constructs are used.

4. On the transfection day, prepare and label the required Eppendorf tubes (two for each construct) and place in a rack under the flow hood. Dry and resuspend the DNA pellet(s) in 10 µl PBS if necessary, and add 250 µl of Invitrogen Optimem 1 medium. Mix gently.

5. In a second tube, mix 12 µl LipofectAMINE 2000 and 250 µl Optimem 1. If more than one construct is used, multiples of the above should be prepared in the same tube and dispensed into aliquots.

6. Add 250 µl of the LipofectAMINE 2000/Optimem mix to each DNA tube, mix and incubate at room temperature for 30 minutes to allow lipoplexes to form.

7. Remove cells from the incubator and aspirate the medium. Add 3 ml of fresh medium (or conditioned medium if necessary) to one well, which will act as the transfection control, and add 2.5 ml of medium to the others.

8. Add the transfection mixes to the wells as appropriate. Swirl the plate to gently mix the reagent with the medium.

9. Incubate overnight at 37°C in a humidified incubator with 5% CO_2.

10. Refresh the medium in the morning, and be aware that there may be a significant amount of cell death depending on the cell line. Incubate overnight.

11. Refresh the medium in the morning and incubate overnight once again.

12. For transient expression experiments, the cells can be assayed the following day. For stable transformation, replace the standard medium with selective medium and refresh daily for 1–2 weeks.

13. At this time, the cells in the control well should be dead and transfected cells should have formed distinct colonies.

Protocol 2.4: Electroporation of mammalian cells

1. Plate approximately 2×10^5 cells in each well of the appropriate number of six-well plates and incubate for 48 hours. Prepare DNA solutions at 1 mg ml^{-1}.

2. Harvest exponentially growing cells by scraping from the wells using a 'rubber policeman'. Pellet the cells at 500 *g* for 5 minutes at 4°C and then resuspend in 1 ml of chilled medium.

3. Determine the cell density using a hemocytometer, then pellet the cells as above and resuspend in PBS or in the electroporation buffer recommended by the instrument supplier at a density of approximately 10^6–10^7 cells per ml. The exact density will vary for different cell lines and the instrument supplier may make recommendations.

4. Transfer aliquots of the cells to prelabeled electroporation cuvettes on ice. The volume required will depend on the electroporator and may vary from 100 μl up to 1 ml.

5. The settings for the electroporator will depend on the cell line being transfected and the device being used. For instruments like the Bio-Rad GenePulser, the optimum parameters will already be known for many cell lines. If this is not the case, try a capacitance between 950 and 1050 μF and test a range of voltages from 200 to 350 V using transient reporter gene assays to establish the best conditions. The critical field strength used with the Eppendorf multiporator is calculated according to a formula in the basic applications manual supplied with the instrument. It is advisable to test voltages ranging from 1× to 5× this value against a number of field lengths (e.g. 50, 75 and 100 μs). If none of these parameters work, it may be necessary to test at higher voltages or with multiple pulses. Some suppliers recommend that the electroporator is discharged through a blank cuvette prior to the actual experiment.

6. Transfer 50 μg of plasmid DNA into the cell suspension in the cuvette and mix carefully, avoiding bubbles.

7. Insert the cuvette in the electroporator and discharge. Leave the cells to stand for 10 minutes at room temperature.

8. Transfer the cells to prewarmed medium and plate out.

9. Incubate overnight and refresh the medium in the morning.

10. The following day, the cells can be harvested for transient assays or refreshed with selective medium to isolate stable transformants.

Cell-mediated gene transfer

3

3.1 Introduction

Conjugal gene transfer between bacteria was discovered in 1946 (1) and was, for a long time, thought to occur only between prokaryotic cells. In the early 1970s, however, it was established that crown gall disease, an affliction of many broad-leaf plants, was also the result of bacterial gene transfer (2). In this case, a section of DNA was transferred from a resident plasmid present in virulent strains of the soil-dwelling bacterium *Agrobacterium tumefaciens* into the genome of wounded plant cells. Other *Agrobacterium* species were soon discovered which could transfer DNA to plants, and the mechanism was shown to be broadly similar to bacterial conjugation (and likewise dependent on the presence of a specific virulence plasmid). To this day, *Agrobacterium*-mediated transformation remains one of the most widely used methods for the genetic manipulation of plants, and many vectors have been derived from the Ti (tumor-inducing) plasmids of virulent *A. tumefaciens* strains. In 1989, Heinemann and Sprague presented a ground-breaking *Nature* paper in which they showed that bacteria could also transfer plasmid DNA to yeast (3). The existence of natural mechanisms for the active transfer of DNA from bacteria to yeast and to plants suggested that active gene transfer from bacteria to animal cells should also be possible.

In 1980, Schaffner described the transformation of cultured mammalian cells by fusion with plasmid-laden protoplasts of the bacterium *Escherichia coli* (4). This technique was the first demonstration of gene transfer between bacterial and mammalian cells, but it can be regarded essentially as a chemical transfection method because the transfer mechanism was passive. That is, the cells were merely vehicles for the DNA and gene transfer was only achieved under highly artificial circumstances – the cells were treated with chloramphenicol to amplify the plasmid copy number at the expense of the bacterial genome, the cell wall was removed, and polyethylene glycol (PEG) was added to promote fusion between the donor and recipient cells when protoplasts were centrifuged onto the mammalian recipients. Under natural conditions, gene transfer would not occur in this manner. Similar protocols have been published in which passive gene transfer to mammalian cells was achieved by fusion to the hemoglobin-free ghosts of erythrocytes preloaded with plasmid DNA. More recently, specialized techniques have been developed for the transfer of large DNA clones (such as YACs) and even whole chromosomes. These involve the fusion of mammalian cells with yeast spheroplasts (the equivalent of bacterial protoplasts) or so-called microcells derived by cell fragmentation and the rescue of individual chromosomes within lipid vesicles. These techniques are discussed in more detail at the end of this chapter.

The first report of active gene transfer from living bacteria to mammalian cells was published in the journal *Science* in 1995 (5). The paper described the use of *Shigella flexneri*, the pathogen responsible for dysentery, for the delivery of plasmid DNA in mice. This and other early reports were greeted with a fair amount of skepticism, but as further examples were published, this new gene transfer strategy (sometimes described as *bactofection*) became accepted as a useful and efficient new methodology. Unlike the conjugation-type and fusion-type mechanisms discussed above, this new wave of cell-mediated gene transfer methods resulted from the invasion of animal cells by bacteria, followed by the lysis of the bacterial cells and the release of plasmid DNA (6–9). Although the majority of bacterial gene transfer vectors for animal cells use this invasive principle, Kunik and colleagues demonstrated, in 2001, that human HeLa cells could be transformed by *A. tumefaciens* (10). This showed that the conjugal gene transfer process could also involve animal cells under the appropriate culture conditions. At least six bacterial vectors for the transformation of animal cells have now been described, and we consider them in turn. (See references (1–10).)

3.2 Cells as active gene transfer vectors

Shigella flexneri as a gene transfer vector

As stated above, *Shigella flexneri* was the first bacterial species shown to transfer DNA actively into animal cells. Wild-type strains of the bacterium invade host cells, typically those of the gut epithelium, by inducing phagocytosis. They then escape from the phagocytic vacuole to multiply in the cytoplasm, eventually causing host cell lysis and death (*Figure 3.1*). All these functions – induction of phagocytosis, escape from the vesicle and rapid intracellular proliferation – are encoded on a large virulence plasmid. The exploitation of and subsequent escape from the phagocytic uptake pathway shows that the bacteria use a parallel and broadly similar method to viruses in order to gain access to the cell.

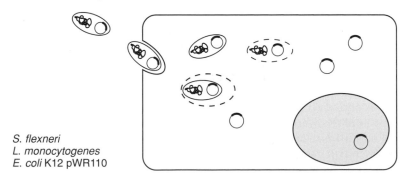

S. flexneri
L. monocytogenes
E. coli K12 pWR110

Figure 3.1

Active gene transfer from *S. flexneri* and *L. monocytogenes* to animal cells involves invasion via the stimulation of phagocytosis followed by escape from the phagocytic vacuole and the release of plasmid DNA into the cytoplasm, where some is transported to the nucleus. Reprinted from Current Opinion in Biotechnology, Vol 12, Weiss and Chakraborty, "*Tranfer of eukaryotic . . .*," Pages 467–472, Copyright (2001), with permission from Elsevier.

To achieve gene transfer without killing the host cells, strains of *S. flexneri* have been developed which are attenuated by mutation, making them metabolically dependent on the host cell. For example, strains have been generated that are deficient in one of the essential steps of cell wall synthesis, or that lack the metabolic capability to synthesize particular nucleotides or amino acids. Attenuated cells proliferate normally under permissive conditions, but if the necessary metabolites are withdrawn then the bacteria die, releasing their cargo of plasmid DNA into the host cell's cytoplasm. As in chemical transfection methods, some of this plasmid DNA makes its way to the nucleus via an intrinsic transport pathway and can be expressed, although the mechanism is poorly characterized. *S. flexneri* has been used as a delivery vector both for cultured cells and for in vivo applications. Indeed, the first use of this vector was for the nasal delivery of DNA vaccines against the hemagglutinin and nucleocapsid proteins of measles virus in mice, resulting in the activation of T-cells and an increase in antibody titers. Similar responses have been shown in mice bactofected with other antigen-encoding sequences, such as tetanus toxin genes and the *E. coli* gene encoding β-galactosidase. (See references (5,11).)

Salmonella typhimurium as a gene transfer vector

While *Shigella flexneri* was the first intact bacterial vector used for mammalian cells, *Salmonella typhimurium* has been used the most extensively. Like *S. flexneri*, *S. typhimurium* is an invasive pathogen, but it employs a slightly different infection strategy, which involves replication *within* the phagocytic vacuole (*Figure 3.2*). As discussed above, preservation of the host cell requires vector attenuation, and in the case of *S. typhimurium* has again been achieved through the use of strains defective in amino acid or nucleotide metabolism. Such attenuated mutants are destroyed in the phagocytic vesicle under restrictive conditions, releasing the cargo of plasmid DNA. This DNA must escape from the phagocytic vesicle and make its way to the nucleus for transformation to be successful.

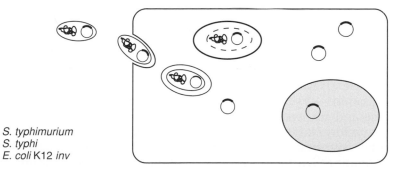

S. typhimurium
S. typhi
E. coli K12 *inv*

Figure 3.2

Active gene transfer from *Salmonella* species to animal cells involves invasion via the stimulation of phagocytosis followed by the release of plasmid DNA within the phagocytic vacuole. Some of this DNA escapes into the cytoplasm and finds its way to the nucleus. Reprinted from Current Opinion in Biotechnology, Vol 12, Weiss and Chakraborty, "*Tranfer of eukaryotic . . .,*" Pages 467–472, Copyright (2001), with permission from Elsevier.

S. typhimurium has been used to express a number of different antigens, including *Listeria monocytogenes* virulence factors, the *Chlamydia trachomatis* major outer membrane protein and various tumor autoantigens. It has also been used to express the markers β-galactosidase and green fluorescent protein, ovalbumin, various cytokines and the pore-forming toxin listeriolysin. The latter is very interesting because the expression of this protein allows *S. typhimurium* to escape from the phagocytic vacuole and therefore mimic the infection strategy of *S. flexneri* and *L. monocytogenes*. In several experiments, this has been shown to enhance the expression of other transgenes, including reporter genes and antigen-encoding genes, presumably by protecting the released DNA from degradation.

The major application of *S. typhimurium* as a gene transfer vector has been to deliver vaccines into mice. Antibodies have been detected against all of the antigens listed above following either oral or nasal inoculation with recombinant *Salmonella*, and the presence of cytotoxic and helper T cells has been reported. *Salmonella*-mediated gene delivery has been used with success to retard tumor development: tumors in mice were induced by the murine melanoma line B16, but their growth was restricted by *Salmonella* vectors carrying epitopes of the autologous tumor antigens gp100 and TRP2 fused to ubiquitin, suggesting that under such circumstances peripheral tolerance towards autologous antigens can be overcome. In a murine B-cell lymphoma system, stimulation via CD40L leads to growth suppression in vitro and in vivo. Oral administration of *Salmonella* carrying expression plasmids encoding a soluble form of the human CD40L protected mice against a simultaneous challenge with the B-cell lymphoma. Human CD40L was detectable for several weeks in the serum of these mice. Similarly, mice defective in interferon-γ (IFNγ) that would normally succumb to a challenge with the metabolically dependent *aroA* strain of *S. typhimurium* were protected when these bacteria carried an expression plasmid encoding IFNγ. So far, DNA transfer using *S. typhimurium aroA* has only been achieved with primary macrophages, but strains secreting listeriolysin as discussed above will help to address this limitation. (See references (12–14).)

Salmonella typhi as a gene transfer vector

Salmonella typhi is the agent responsible for typhoid fever, and like *S. typhimurium* it invades host cells but remains trapped within the phagocytic vesicle (*Figure 3.2*). The *S. typhi* strain most commonly used for immunization — Ty21a *galE* — lacks an essential component for cell-wall synthesis and is therefore naturally attenuated, allowing its use as a gene transfer vector. Further strains have been developed which are deficient in nucleotide synthesis, including the improved attenuated strain *gua*BA, which cannot synthesize guanine. Administering such strains either intraperitoneally or nasally to mice resulted in cytotoxic T-cell responses against the nucleocapsid protein of measles virus and gp140 of human immunodeficiency virus. A comprehensive study was performed using tetanus toxin as the antigen to compare nasally delivered *S. typhi* CVD915 Δ*gua*BA carrying either a eukaryotic expression plasmid or a plasmid with a prokaryotic intracellular inducible promoter. Antibody responses elicited by both immunizations were better than those elicited by naked DNA, and bacteria carrying the eukaryotic expression plasmid induced higher antibody levels than those carrying the prokaryotic expression plasmid. A booster reaction was observed using either of the two recombinant bacteria despite the presence of antibodies against lipopolysaccharide after the primary immunization. This suggests that *Salmonella*-mediated DNA vaccination should also be possible in individuals that have previously encountered *Salmonella*. (See reference (15).)

Listeria monocytogenes as a gene transfer vector

Listeria monocytogenes is the causative agent of listeria. Its invasive infection strategy is similar to that of *S. flexneri* in that it escapes from the phagocytic vacuole and replicates in the cytoplasm. Once in the cytoplasm, the bacterium becomes motile by recruiting components of the host cell cytoskeleton and subsequently spreads to neighboring cells. Although there are no available auxotrophic strains of *L. monocytogenes* that can invade host cells efficiently, wild-type strains can be used as delivery vectors. Also, a strain has been engineered to contain a suicide gene, encoding an autolysin that is activated inside the cell. Using these bacteria, several reporter genes have been introduced into a murine macrophage cell line and into primary human dendritic cells. Bacteria carrying a GFP-encoding plasmid have also been used for *in vivo* transfer, following injection into the peritoneum of rodents. Cells harvested from these animals after a few days yielded macrophages that expressed the reporter gene. An alternative strategy for *in vitro* experiments is to add antibiotics to the culture medium after an appropriate infection time, killing the bacteria and releasing plasmid DNA into the cells. Several different cell types have been bactofected in this manner. With some cell lines, transfer to more then 10% of cells could be achieved and stable transgene integration rates ranged from 10^{-7} to 10^{-2}. (See references (16–20).)

Escherichia coli as a gene transfer vector

E. coli is not normally an invasive species, but it can be converted into an invasive pathogen by transformation with virulence genes or virulence plasmids from other bacteria. The standard K12 laboratory strain of *E. coli* has been attenuated by mutating the *dapB* gene, which is required for purine synthesis, and transformed with the virulence plasmid from *S. flexneri*. This confers on the *E. coli* strain the ability to invade mammalian cells and escape from the phagocytic vesicle before lysis, resulting in the successful transfer of plasmid DNA to the mammalian host. *E. coli* has also been transformed with the invasin gene from *Yersinia pseudotuberculosis*, which allowed the bacteria to invade the host cell but not escape from the phagocytic vesicle, thus mimicking the infection strategy of *Salmonella* vectors. DNA transfer to the host was also observed in these experiments. (See reference (21).)

Agrobacterium tumefaciens as a gene transfer vector for animal cells

As stated at the beginning of this chapter, *Agrobacterium tumefaciens* is widely exploited as a gene transfer vector for plants, but has also been shown to transfer DNA to human cell lines. The gene transfer mechanism is not invasive (*Figure 3.3*). The bacteria attach to human cells as they do to plant cells and DNA transfer is dependent on the same transfer apparatus used in plants. Subsequent analysis using the polymerase chain reaction and DNA sequencing revealed that plasmids had integrated correctly into random genomic sites in HeLa cells. (See reference (10).)

3.3   Cells as passive transfer vehicles for gene transfer

Bacterial protoplasts as gene transfer vehicles

As discussed in the introduction, bacterial cells can be used not only for the active transfer of DNA to animal cells, but also as passive vehicles. In this way, their role in the gene transfer process is similar to that of liposomes (p. 23) in that they deliver the DNA by fusing with the plasma membrane of the recipient cell, depositing the plasmid DNA into the cytoplasm. In the original description of the technique (4), Schaffner grew *E. coli* in standard medium supplemented with 250 µg ml^{-1} of chloramphenicol to amplify the plasmid DNA. The bacterial cells were then recovered by centrifugation, resuspended in ice-cold sucrose solution and treated with lysozyme for 5–10 minutes to digest away the cell wall, leaving behind fragile bacterial protoplasts. After making up a suspension containing 10^8–10^9 protoplasts per ml (equivalent to 10 000 protoplasts per mammalian cell), the medium was removed from the mammalian cells and the protoplasts were added. This step was followed by a brief centrifugation to bring the donor and recipient cells into contact, and then the addition of 50% PEG to stimulate membrane fusion. For transient expression experiments, the cells were harvested about 24 hours later, while for stable transfection the mammalian cells were allowed to grow in non-selective medium for 24

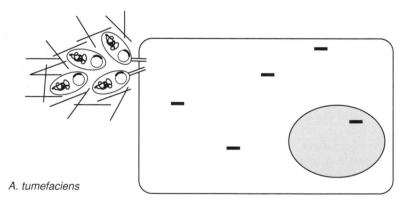

A. tumefaciens

Figure 3.3

Agrobacterium tumefaciens uses a non-invasive gene transfer strategy, in which cells attached to the outside of the host deliver a linear fragment of DNA derived from a resident plasmid. The machinery required for the preparation of this transferred DNA (T-DNA) and its delivery is encoded by a virulence plasmid in the bacteria. Reprinted from Current Opinion in Biotechnology; Vol 12, Weiss and Chakraborty, "*Tranfer of eukaryotic . . .*," Pages 467–472, Copyright (2001), with permission from Elsevier.

hours allowing the transgenes time to integrate and initiate expression, before transfer to selective medium. Overall, the transient transfection efficiency ranges from about 5% to 100% depending on the bacterial strain used as the vector, and the stable transformation efficiency approaches 0.02%. The technique is not widely used anymore because it is rather laborious, and has been supplanted by the more rapid and efficient methods based on cationic lipids and polymers (p. 24). (See reference (4).)

Yeast spheroplasts as gene transfer vehicles

The rationale of using bacterial protoplasts as gene transfer vehicles is that plasmids are cloned and maintained in the bacteria, so this direct procedure avoids having to extract and purify the plasmid DNA. Although this would not be regarded as a problem today, with very rapid and efficient methods for plasmid purification, the same cannot be said for the much larger YAC vectors which can be up to 2 Mb in length and are extremely susceptible to shear forces. Therefore, an analogous procedure has been developed for the transformation of mammalian cells with yeast spheroplasts, the functional equivalent of bacterial protoplasts, for the direct transfer of YAC vector DNA. The following protocol from Allshire and colleagues is very similar to Schaffner's protocol for bacterial protoplasts (22). Yeast cells are grown in standard YPD medium, pelleted, resuspended in a sorbitol solution to represent the intracellular osmolarity and then treated with zymolyase for 20 min to remove the cell wall. The recipient mammalian cells are then trypsinized to remove them from the substrate, washed, resuspended in fresh medium and pelleted, and the yeast spheroplast suspension is layered on the top. The mixture of cells is then centrifuged to bring the donors and recipients into contact, gently resuspended in medium and mixed with 40% PEG to induce fusion. After 24 hours, the mammalian cells are replated in a medium which selects for the marker gene carried on the YAC vector. (See references (22,23).)

Mammalian cells as gene transfer vehicles

Direct passive gene transfer between mammalian cells has a long and successful history, which can be traced back to the development of somatic hybridization technology via cell fusion. One application of this technique is the use of interspecies somatic cell hybrids for gene mapping. In the case of the human genome, this is achieved by hybridizing human and rodent cells in culture by mixing the co-cultured cells with PEG. The resulting hybrid cell is known as a heterokaryon because it contains two nuclei (one human, one rodent). However, after a while the nuclei divide, which involves the breakdown of the nuclear membrane. In many cases, the chromosomes mingle and the new nuclear membrane forms around the combined hybrid genome. This state of affairs is unstable, and for some unknown reason the human chromosomes are prevented from replicating, and are lost as the hybrid cell proliferates. The result is a series of clones carrying individual human chromosomes, and it is possible to generate hybrid cell panels in which each human chromosome is individually represented. These are known as monochromosomal hybrids (*Figure 3.4*). A more

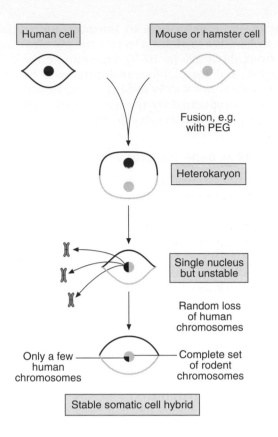

Figure 3.4

The principle of somatic cell hybridization, resulting in panels of hybrid cells containing a complete complement of rodent chromosomes and only a few human chromosomes.

systematic way to generate monochromosomal hybrids is the technique of microcell-mediated gene transfer. In this method, human cells are treated with colcemid, which interferes with nuclear division and causes the nucleus to fragment such that groups of chromosomes and individual chromosomes become partitioned within their own nuclear envelopes (micronuclei). The centrifugation of such cells can result in the formation of microcells, where a micronucleus containing a single chromosome is enclosed within a small ring of cytoplasm. These microcells can be recovered and induced to fuse with rodent cells, generating defined monochromosomal hybrids. One application of this approach is the location of disease genes which display a distinct cellular phenotype (*Figure 3.5*). If the human cells are subjected to a lethal radiation dose prior to fusion, their chromosomes will contain a number of breaks, proportional to the dose of radioactivity. In this way, fusion followed by the selective loss of human chromosomes results in hybrid cell lines carrying small fragments of the human genome, permitting much finer mapping. Such cell lines are called radiation hybrids. (See reference (24).)

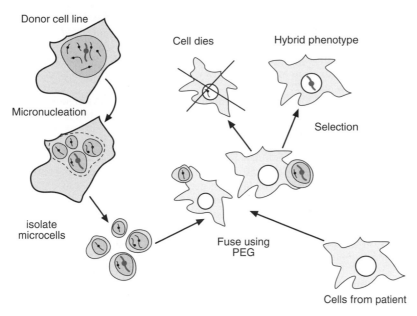

Donor cell line

Cell dies

Hybrid phenotype

Micronucleation

Selection

isolate
microcells

Fuse using
PEG

Cells from patient

Figure 3.5

Microcell-mediated gene transfer for the mapping of disease genes. Treatment of human cells with colcemid induces micronucleation, which can result in the recovery of microcells. In this application, the human cells have been transfected with a selectable marker which has integrated into one particular chromosome. Microcells carrying this marker are fused to patient cells representing a specific genetic disease whose locus is unknown. Hybrid cells surviving selection contain the known chromosome. If the genetic disease has a cellular phenotype, such as metabolic dependency or inefficient DNA repair, the surviving monochromosomal hybrids can be tested for their phenotype. If the phenotype has been corrected, the disease gene is likely to be located on the rescued chromosome.

References

1. Lederberg J, Tatum EL (1946) Gene recombination in *E. coli*. *Nature* **158**: 558.
2. Gelvin SB (2003) *Agrobacterium*-mediated plant transformation: the biology behind the "gene-jockeying" tool. *Microbiol Mol Biol Rev* **67**: 13–37.
3. Heinemann JA, Sprague GF Jr (1989) Bacterial conjugative plasmids mobilize DNA transfer between bacteria and yeast. *Nature* **340**: 205–209.
4. Schaffner W (1980) Direct transfer of cloned genes from bacteria to mammalian cells. *Proc Natl Acad Sci USA* **77**: 2163–2167.
5. Sizemore DR, Branstrom AA, Sadoff JC (1995) Attenuated shigella as a DNA delivery vehicle for DNA-mediated immunization. *Science* **270**: 299–302.
6. Lowrie DB (1998) DNA vaccination exploits normal biology. *Nat Med* **4**: 147–148.
7. Higgins DE, Portnoy DA (1998) Bacterial delivery of DNA evolves. *Nat Biotechnol* **16**: 138–139.
8. Weiss S, Chakraborty T (2001) Transfer of eukaryotic expression plasmids to mammalian host cells by bacterial carriers. *Curr Opin Biotechnol* **2**: 467–472.
9. Shata MT, Stevceva L, Agwale S, Lewis GK, Hone DM (2000) Recent advances with recombinant bacterial vaccine vectors. *Mol Med Today* **6**: 66–71.

10. Kunik T, Tzfira T, Kapulnik Y, Gafni Y, Dingwall C, Citovsky V (2001) Genetic transformation of HeLa cells by *Agrobacterium*. *Proc Natl Acad Sci USA* **98**: 1871–1876.
11. Fennelly GJ, Khan SA, Abadi MA, Wild TF, Bloom BR (1999) Mucosal DNA vaccine immunization against measles with a highly attenuated *Shigella flexneri* vector. *J Immunol* **162**: 1603–1610.
12. Xiang R, Lode HN, Chao TH, Ruehlmann JM, Dolman CS, Rodriguez F, Whitton JL, Overwijk WW, Restifo NP, Reisfeld RA (2000) An autologous oral DNA vaccine protects against murine melanoma. *Proc Natl Acad Sci USA* **97**: 5492–5497.
13. Paglia P, Terrazzini N, Schulze K, Guzman CA, Colombo MP (2000) In vivo correction of genetic defects of monocyte/macrophages using attenuated *Salmonella* as oral vectors for targeted gene delivery. *Gene Ther* **7**: 1725–1730.
14. Darji A, zur Lage S, Garbe AI, Chakraborty T, Weiss S (2000) Oral delivery of DNA vaccines using attenuated *Salmonella typhimurium* as carrier. *FEMS Immunol Med Microbiol* **27**: 341–349.
15. Pasetti MF, Anderson RJ, Noriega FR, Levine MM, Sztein MB (1999) Attenuated ΔguaBA *Salmonella typhi* vaccine strain CVD 915 as a live vector utilizing prokaryotic or eukaryotic expression systems to deliver foreign antigens and elicit immune responses. *Clin Immunol* **92**: 76–89.
16. Hense M, Domann E, Krusch S, Wachholz P, Dittmar KEJ, Rohde M, Wehland J, Chakraborty T, Weiss S (2001) Eukaryotic expression plasmid transfer from the intracellular bacterium *Listeria monocytogenes* to host cells. *Cell Microbiol* **3**: 599–609.
17. Pilgrim S, Stritzker J, Schoen C, Kolb-Maurer A, Geginat G, Loessner MJ, Gentschev I, Goebel W (2003) Bactofection of mammalian cells by *Listeria monocytogenes*: improvement and mechanism of DNA delivery. *Gene Ther* **10**: 2036–2045.
18. Grillot-Courvalin C, Goussard S, Courvalin P (2002) Wild-type intracellular bacteria deliver DNA into mammalian cells. *Cell Microbiol* **4**: 177–186.
19. Dietrich G, Bubert A, Gentschev I, Sokolovic Z, Simm A, Catic A, Kaufmann SH, Hess J, Szalay AA, Goebel W (1998) Delivery of antigen-encoding plasmid DNA into the cytosol of macrophages by attenuated suicide *Listeria monocytogenes*. *Nat Biotechnol* **16**: 181–185.
20. Hense M, Domann E, Krusch S, Wachholz P, Dittmar KE, Rohde M, Wehland J, Chakraborty T, Weiss S (2001) Eukaryotic expression plasmid transfer from the intracellular bacterium *Listeria monocytogenes* to host cells. *Cell Microbiol* **3**: 599–609.
21. Grillot-Courvalin C, Goussard S, Huetz F, Ojcius DM, Courvalin P (1998) Functional gene transfer from intracellular bacteria to mammalian cells. *Nat Biotechnol* **16**: 862–866.
22. Allshire RC, Cranston G, Gosden JR, Maule JC, Hastie ND, Fantes PA (1987) A fission yeast chromosome can replicate autonomously in mouse cells. *Cell* **50**: 391–403.
23. Simpson K, McGuigan A, Huxley C (1996) Stable episomal maintenance of yeast artificial chromosomes in human cells. *Mol Cell Biol* **16**: 5117–5126.
24. Doherty AMO, Fisher EMC (2003) Microcell-mediated chromosome transfer (MMCT): small cells with huge potential. *Mamm Genome* **14**: 583–592.

Gene transfer using viral vectors

4

4.1   Introduction

General properties of viral vectors

Viruses have evolved to deliver nucleic acids safely into animal cells. They do this first by binding to specific receptors on the cell surface, and then (in the case of enveloped viruses) by either fusing directly with the plasma membrane or, following uptake by endocytosis, by fusing to the endosomal membrane. In contrast, non-enveloped viruses penetrate or disrupt the plasma or endosomal membranes with specific virion proteins (*Figure 4.1*). In each case, the viral genome ends up being uncoated in the cytoplasm and transported to its normal replication site, which for many but not all viruses is the nucleus.

The transfer of exogenous DNA (or RNA) into animal cells as part of a recombinant viral particle is known as *transduction*. The advantages of viral transduction methods over transfection with naked plasmid DNA include the high efficiency of gene transfer (because a natural delivery process is used), the encapsulation of the transgene within a protective protein coat prior to and during internalization, the tendency of many viruses to block host cell protein synthesis and favor the expression of viral genes (including transduced foreign genes), and the availability of viruses with broad or narrow cellular host ranges as appropriate for the aims of the experiment. Furthermore, different viruses can achieve different transformation objectives: short-term infections with high-level transient expression (e.g. adenoviruses), long-term expression by maintaining the genome as a latent episomal replicon (e.g. herpesviruses), or stable integration of DNA into the host cell genome (e.g. retroviruses). The integrated form of a viral genome is known as a *provirus*, and the same term is sometimes used to describe the plasmid-like episomal form of a latent virus genome. Many different animal viruses have been exploited as vectors, mostly for foreign gene expression in cultured cells but also for gene transfer *in vivo*. Eight types of viral vector have been used in phase I and in some cases phase II clinical gene therapy trials: adenovirus, adeno-associated virus, herpes simplex virus, oncoretroviruses, lentivirus (including HIV), poxviruses (including Vaccinia) and – each on one occasion thus far – Semliki Forest virus and measles virus. There have also been some combined trials, e.g. using adenovirus and retrovirus vectors. Properties of viruses used as gene transfer vectors are summarized in *Table 4.1*. Disadvantages of viruses as gene transfer vectors include the more complex cloning strategies compared to plasmid vectors, and (for *in vivo* transfer) health risks resulting from the possible production of infectious wild-type viruses during vector construction and allergic reactions to the vector following viral administration.

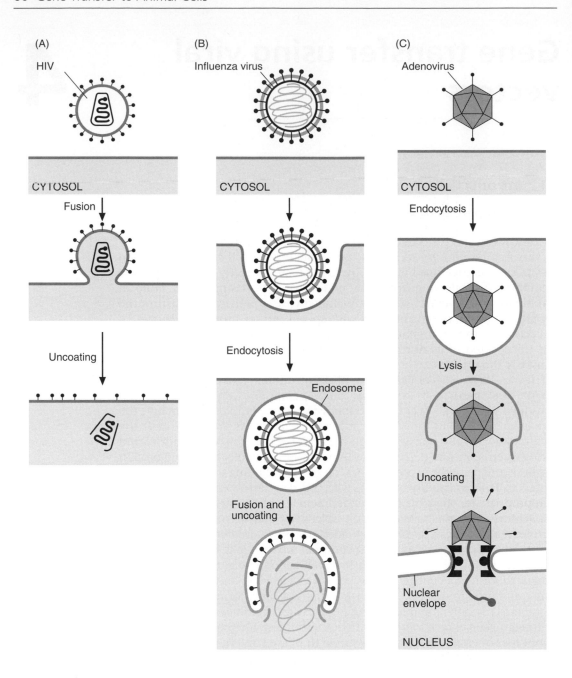

Figure 4.1

Three virus strategies for uncoating. (A) An enveloped virus (e.g. HIV) fuses directly with the host cell plasma membrane, releasing its capsid into the cytosol. (B) An enveloped virus (e.g. influenza virus) binds to cell-surface receptors which triggers receptor-mediated endocytosis. The virus envelope fuses with the endosomal membrane when the endosome acidifies, releasing the nucleocapsid into the cytosol. (C) A non-enveloped virus (e.g. adenovirus) induces receptor-mediated endocytosis and disrupts the endosomal membrane, to release part of the capsid into the cytosol. The capsid docks onto a nuclear pore and releases its DNA genome directly into the nucleus.

Table 4.1 Properties of major viral vector classes used in gene transfer

Virus	Genome	Fate in cell	Host	Applications
Adenovirus	dsDNA, 36 kbp	Nuclear, episomal	Many mammalian	Transient expression in cells, short-term gene therapy
Adeno-associated virus	ssDNA, 5kb	Nuclear, integrates	Many mammalian	Stable transformation of cells, long-term gene therapy
Alphavirus	ssRNA, 10 kb	Cytoplasmic	Wide host range	Transient expression in cells and *in vivo*
Baculovirus	dsDNA, 100 kbp	Nuclear, episomal	Insect cells (Latent in mammalian cells)	High level transient expression in insects, stable episomal expression in mammals
Herpes simplex	dsDNA, 100 kbp	Nuclear episomal	Wide host range for lytic infections, latency in neurons	Transient expression in cells and *in vivo*, long term expression in neurons
Lentiviruses	ssRNA, 7 kb	Nuclear, integrates	HIV restricted to human immune cells with CD4 receptor	Stable transformation of cells, long-term gene therapy, transgenesis
Poxviruses incl. vaccinia	dsDNA, 300 kbp	Cytoplasmic	Wide host range	Transient expression in cells and *in vivo*, vaccine development
Retrovirus (MLV-type)	ssRNA, 5 kb	Nuclear, integrates	Many mammals and birds Dividing cells only	Stable transformation of cells, long-term gene therapy, transgenesis

General strategies for viral vector construction

Like foreign genes integrated in cellular genomes, foreign genes inserted into a viral genome are known as *transgenes*. There are three strategies for the incorporation of foreign genes into viral genomes (*Figure 4.2*). In two of these strategies, the resulting virus is described as *helper-independent*, meaning that the recombinant virus is able to propagate normally even though it carries foreign DNA. In one approach, foreign DNA can be added to the entire viral genome without loss of viral genomic sequences. Vectors designed on this principle are termed *insertion vectors*, and typically the foreign DNA is inserted between functional viral genes, or at the edge of a linear genome. In the other approach, foreign DNA can be used to replace a non-essential viral gene (non-essential, at least, for productive infection in the particular system used in the experiment; for example, a gene that is necessary for infection *in vivo* may not be required for viral propagation in cell lines). Vectors designed on this principle are termed *replacement vectors*, and the wild-type viral DNA fragment that is replaced is termed the *stuffer fragment*. The third strategy also involves replacement, but in this case one or more essential viral genes are replaced with foreign DNA. This renders the recombinant vector *helper-dependent*, and missing essential functions must be supplied from another source. The missing genes may be required for replication, viral gene expression, packaging and assembly or other functions. The missing functions may be supplied using a *helper virus* (a virus carrying the essential functions, which is co-introduced with the vector) or using a *helper plasmid*, which is transfected into the cells infected with the vector. Alternatively, dedicated cell lines sometimes termed *packaging lines* or *complementary lines* can be used to propagate the vector (*Figure 4.2*). These cells are transformed with a deficient viral genome that is itself incapable of replication, but supplies helper functions in *trans* to the vector. The extreme form of this strategy uses derivatives of the viral genome in which all viral coding sequences are deleted, leaving behind only those *cis*-acting elements required for replication, packaging and/or proviral integration – these have various names: *amplicons*, *gutted* (or *gutless*) *vectors*, or *fully deleted (FD) vectors*. The choice of strategy depends on many factors including genome size, the availability of packaging lines, the number and nature of non-essential genes, and the packaging capacity of the viral capsid. Many viruses (e.g. papovaviruses) package DNA into a preformed capsid and thus have a strictly defined packaging capacity that limits the size of foreign DNA. Others (e.g. baculoviruses) form the capsid around the genome, and no such limitation exists. An advantage of using packaging lines or other helper systems with enveloped viruses is that the host range of the virus can be specified by *pseudotyping*. This means that the envelope from one virus can be used to package a different virus, allowing the recombinant virus to interact with the receptors usually recognized by the envelope donor. The host range of the recombinant vector can therefore be defined by the helper system.

There are several general techniques for placing foreign DNA into viral genomes. The two most popular are ligation and homologous recombination (*Figure 4.3*). In the ligation strategy, restriction enzymes are used to prepare vector and foreign DNA and the two elements are joined *in vitro*

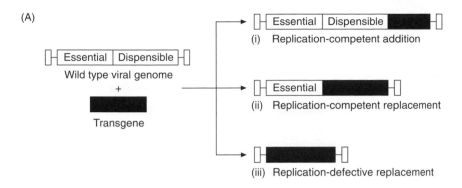

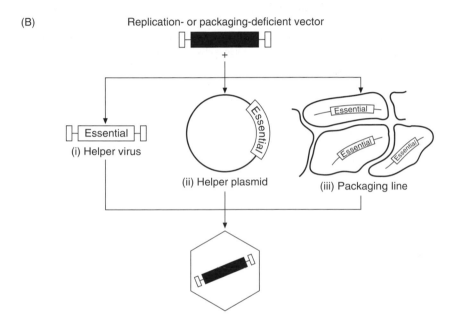

Figure 4.2

(A) Strategies for viral vector construction. (i) The transgene can be added to the entire viral genome. (ii) The transgene can be used to replace a non-essential part of the viral genome. Both the above vectors are replication and packaging competent. (iii) The transgene can replace an essential part of the viral genome rendering the vector replication or packaging deficient. (B) In the last case, the replication or packaging function must be supplied in trans from (i) a helper virus, (ii) a helper plasmid transfected into the infected cells, or (iii) a permanently transformed helper cell line.

using DNA ligase. In the homologous recombination strategy, a plasmid carrying the foreign DNA within a viral homology region is transfected into cells infected with the parental vector, and recombinants are generated by crossing-over within the homology region. The homologous recombination strategy is favored for viruses with large genomes, which are difficult to manipulate directly (e.g. baculoviruses, herpesviruses, poxviruses). The use

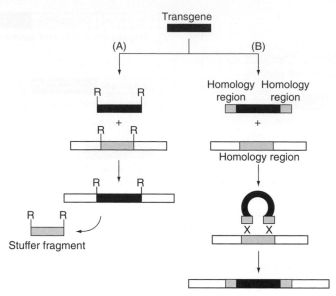

Figure 4.3

Transgene insertion into viral vectors. (A) The transgene inserts at a single site (not shown) or is used to replace a stuffer fragment – this process occurs *in vitro* and requires DNA ligase. (B) The transgene is presented within a viral homology region and replaces the matching sequence in the viral genome. This process occurs inside the infected cell.

of helper viruses that are themselves replication-defective or that carry mutations allowing counterselection is useful for preparing pure stocks of recombinant vector. Contemporary systems are also designed so that two or more entirely independent recombination events are required to generate replication-competent, wild-type virus genomes. Novel strategies involving site-specific recombination and transposition have also been used to generate recombinant viruses.

4.2 Gene delivery with adenovirus vectors

Overview

The adenoviruses (*Adenoviridae*) are non-enveloped DNA viruses, which possess an icosahedral capsid about 100 nm in diameter containing a double-stranded, linear genome approximately 36 kbp in length. Of the two known genera, only the mastadenoviruses can infect mammalian cells. Six subgroups of human adenoviruses have been defined (named A–F) based on cellular agglutination assays and genomic GC content. Over 50 distinct serotypes of human adenovirus have been isolated, but most adenovirus vectors are derivatives of subgroup C serotypes Ad2 and Ad5. These are well known for causing mild upper respiratory tract diseases in humans, but they also infect other mammals and display a broad cellular host range.

The major advantages of adenoviral vectors are that they can be purified to extremely high titers (10^{12}–10^{13} particles per ml), which makes them highly suited for *in vivo* applications, and the efficiency of gene transfer approaches 100% if the target cells bear the appropriate receptors. Adenovirus vectors have a broad species and cellular host range (including both dividing and postmitotic cells) and they are relatively easy to manipulate *in vitro*. Adenovirus replacement vectors in common use can accommodate up to 7.5 kbp of foreign DNA, although some fully deleted vectors for gene therapy have been designed to accommodate the theoretical maximum of approximately 30 kbp. Adenoviral vectors show a low efficiency of stable transformation, so they are generally suitable for transient expression *in vitro* and *in vivo* but they are not useful for the production of stably transformed cell lines. Another disadvantage is their tendency to provoke an immune response *in vivo*, which has led to severe inflammatory responses and at least one death during clinical gene therapy trials. Active research to overcome these problems is ongoing and some recent developments are discussed below.

Adenovirus molecular biology

The adenoviral capsid has 20 facets, each comprising 12 copies of a trimeric hexon protein (*Figure 4.4*). At each vertex, there is a pentameric penton base supporting a trimeric fiber protein ending in a globular knob. These are the major structural components of the capsid and they are all involved in the infection process. Other structural proteins are required for further stabilization of the capsid and the formation of the genomic core.

The infection of target cells occurs in two phases, mediated by two receptors (*Figure 4.5*). First, the globular knob domain of the adenovirus fiber protein interacts with the primary cellular receptor, known as CAR (coxsackievirus and adenovirus receptor). Receptor-mediated endocytosis then internalizes the virion, and this process is enhanced in a second stage by interactions between the penton base protein and integrins αvβ3 and αvβ5 on the cell surface. The adenovirus fiber proteins have the ability to disrupt the endosome, releasing the naked capsid into the cytoplasm where it is transported to the nuclear membrane. Here, the capsid is disassembled and the viral DNA is passed through nuclear pores into the nucleoplasm, where it is transcribed.

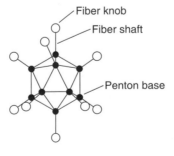

Figure 4.4

Adenoviral capsid structure.

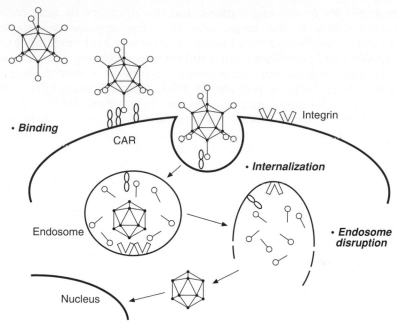

Figure 4.5

Adenoviral infection cycle. Copyright (2004), Humana Press. Used with permissions.

There are approximately ten transcription units in the linear adenovirus genome, but the number of gene products produced is increased by complex patterns of alternative splicing and protein processing. Genes are expressed in two major phases, early genes before replication and late genes coincident with the onset of replication. There are six early transcription units, which from left to right on the adenovirus gene map are designated E1a, E1b, E2a, E2b, E3 and E4 (*Figure 4.6*). Expression of the E1a gene depends only on cellular proteins, and the products are transcriptional regulators of the other early genes. E1b proteins control mRNA export and inhibit apoptosis and host cell protein synthesis. E2a and E2b encode

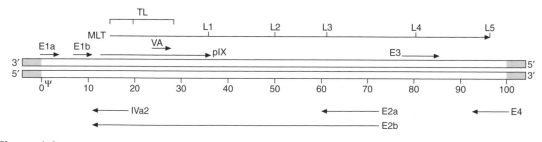

Figure 4.6

Adenoviral genome structure. The genome is conventionally divided into 100 units. Grey boxes = inverted terminal repeats. Arrows represent transcripts, TL = tripartite leader, MLT = major late transcript, Ψ = packaging site.

proteins required for viral DNA replication. E3 proteins help to evade the host immune response by blocking apoptotic pathways, and E4 encodes further transcriptional regulators and modulators of host cell activity. All these genes except E3 are essential for viral propagation in cultured cells.

The early phase of infection takes about 6 hours, after which time progeny viral genomes can be detected. Replication occurs by a displacement mechanism, and begins at inverted repeats at the termini of the viral genome where it is primed by terminal, nucleotide-binding proteins. After replication, early gene expression is switched off and transcription initiates from the single major late promoter, producing five major families of transcripts, L1–L5. All these late transcripts begin with a sequence termed the tripartite leader, an untranslated region containing two introns, which increases the efficiency of protein synthesis. Each transcript family also has a common 3' end, and individual members differ with respect to internal splicing patterns. Further genes within the E1b, E2a and E2b transcription units are expressed during the late phase. There are also several intermediate (or delayed early) genes, e.g. the set of viral associated (VA) RNA genes transcribed by RNA polymerase III. These enhance late viral gene expression by inhibiting DAI protein kinase (dsRNA-activated protein kinase), which otherwise blocks protein synthesis in the cell during late infection (the VA genes are sometimes exploited in transfection experiments – see page 26). Late genes encode viral structural proteins and proteins involved in their processing and assembly to form the mature virions, which are released 20–24 hours after infection.

Construction of recombinant adenovirus vectors

Both replication-competent and replication-deficient adenoviral vectors have been developed (*Figure 4.7*). The replication-competent vectors include E3 replacement vectors and insertion vectors (where foreign DNA is added at the right-hand edge of the linear gene map adjacent to the E4 transcription unit) but these have a low capacity for foreign DNA. The most widely used adenovirus vectors are the E1 replacement vectors, which are replication-defective due to deletion of the E1a and E1b transcription units. These are propagated in a complementary cell line such as the human embryonic kidney cell line HEK 293, which is transformed with the left-most 11% of the adenovirus genome and hence supplies the missing E1a and E1b functions (*Protocol 4.1*). E1 replacement vectors may also lack the E3 transcription unit, which increases the maximum capacity of the vector to approximately 7.5 kbp. The further deletion of the E4 region allows foreign DNA of up to 11 kbp to be inserted, and has the additional advantage of removing adenoviral functions that interfere with host cell physiology. The theoretical maximum capacity of an adenoviral vector is 30 kbp. This requires the deletion of all adenoviral coding DNA, leaving just the *cis*-acting elements required for packaging and replication. Packaging lines like HEK 293 cannot be prepared for adenoviral amplicons because the overexpression of adenoviral late proteins is cytotoxic, so helper viruses or plasmids are required.

Wild-type adenovirus genomes are difficult to manipulate *in vitro* because of their size and the lack of conveniently placed restriction sites. The

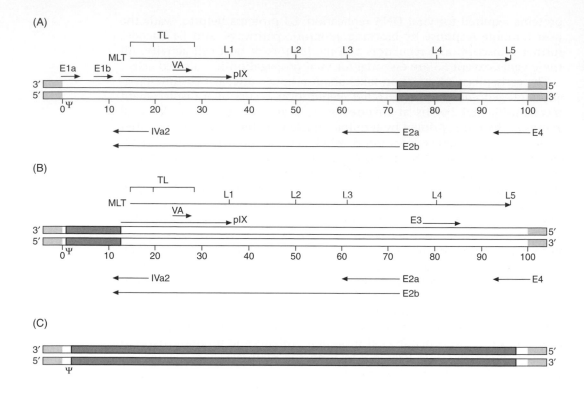

Figure 4.7

Adenovirus replacement vectors, dark bars represent foreign DNA. (A) Replacement of E3 (replication competent). (B) Replacement of E1 (first generation, replication defective; the E3 region may also be deleted to increase capacity). (C) Amplicon vector (second generation, replication defective).

production of recombinant adenoviral vectors by homologous recombination has therefore become a popular strategy, and a series of plasmid vectors has been produced containing adenoviral sequences interrupted by multiple cloning sites to facilitate transgene insertion. Early homology-based vector construction methods involved the transfection of HEK 293 or similar E1-complementing cell lines with two DNA constructs, one comprising the short, leftmost portion of the adenovirus genome with the E1 transcription unit replaced by the transgene, and the other comprising the remaining portion of the genome modified to contain a short homology region matching the 3' end of the transgene to promote recombination. However, this method is rather inefficient because of the low level of homologous recombination in mammalian cells, and there is a further risk that replication-competent viruses could be generated by recombination with the adenovirus helper genes in the genome of the host cells. Therefore, this approach has been replaced by new procedures in which vector construction is carried out in bacterial cells, which are much more amenable to homologous recombination, and the development of helper lines with minimal homology to the recombinant vectors, to avoid the production of replication-competent viruses. The alternative strategy of recombinant vector production *in vitro* has been facilitated by the development of derivatives of the wild-type adenovirus genome containing unique restriction sites for insertion, or paired sites for excision of E1 or E3 stuffer fragments and their replacement with foreign DNA.

Recombinant adenoviral genomes produced as discussed above are infectious and can be introduced into permissive cells by transfection. Evidence of viral infection is seen within a few days, and cell lysates can then be used to infect fresh cells, resulting in the formation of plaques representing viral clones. These can be purified, e.g. by cesium chloride centrifugation, or, as has become more popular more recently, by a combination of ion exchange and reversed-phase chromatography, or immobilized metal-affinity chromatography (IMAC). The viral titer was originally determined using a biological assay and expressed as plaque-forming units (PFU) per ml. Nowadays, it is quicker and more convenient simply to disrupt an aliquot of the virus using sodium dodecylsulfate and determine the optical absorbance of the viral DNA.

Major applications of adenoviral vectors

Gene transfer with adenovirus vectors has two major applications: transient expression of recombinant proteins in cultured cells and gene transfer *in vivo*. For use in cultured cells, the advantages of adenoviral vectors are the broad cellular host range, the high efficiency of gene transfer and the high level of transgene expression (which can result in recombinant protein levels exceeding 30% of total soluble protein). For *in vivo* applications, the advantages are the same but there are also several drawbacks. First, despite the broad host range, not all cells *in vivo* display the required CAR and integrin molecules as receptors, and some cell types that do express these receptors do so at very low levels resulting in a low efficiency of gene transfer. Perhaps the greatest difficulty with these vectors, however, is their toxicity and immunogenicity.

If the intravenous route is used, the proteins of the adenoviral capsid trigger an acute inflammatory response almost immediately after administration, which involves the release of cytokines such as interleukin-8 and the recruitment of immune cells to the liver. Such responses have been noted in several gene therapy trials involving adenoviral vectors, and are considered to be dose-dependent. The death of 18-year-old Jesse Gelsinger during a gene therapy trial for the inherited liver disorder ornithine transcarbamylase deficiency is thought to have been caused by an inflammatory response to the particularly high dose he received, and following this case all gene therapy trials involving adenovirus vectors were halted while safety procedures were examined. One possible approach to overcome the limitation of allergic reactions is the development of vectors that specifically target particular cell types, which would increase gene transfer efficiency and allow initial doses to be reduced. For example, vectors with modified capsids that recognize alternative and more abundant cellular receptors on particular target cell populations, e.g. tumor cells, could be developed.

In the case of E1-deletion vectors, the acute inflammatory response may be followed by a more prolonged cellular immune response as adenoviral peptides are displayed on the surface of infected cells. This is also a major reason for the short duration of transgene expression mediated by adenoviral vectors, as many of the transformed cells are eliminated by the immune system. Finally, a humoral immune response may also occur if antibodies already exist to the vector (as would be the case for repeated dose treatments, for example). These problems can be addressed to a certain extent by the use of the fully deleted amplicon vectors discussed above, which can prolong transgene expression from the typical 2–3 weeks seen with E1-deletion vectors to several months. (See references (1–8).)

4.3   Gene delivery with adeno-associated virus vectors

Overview

Adeno-associated virus (AAV) belongs to the *Dependovirus* subgroup of the *Parvoviridae*, a family of single-stranded DNA viruses with small, non-enveloped icosahedral capsids. There are six serotypes of AAV, and most of the vectors that have been developed are based on AAV2. AAV is naturally replication defective, i.e. it requires the presence of a different virus to act as a helper and provide missing functions that allow the replication cycle to be fulfilled. Adenovirus is a suitable helper virus, hence 'adeno-associated', but herpes virus is also competent to supply the required functions, and to a lesser extent so is Vaccinia virus. The infection cycle of AAV is biphasic. In cells infected with a suitable helper virus, AAV enters the lytic cycle and replicates to a high copy number. In other cells, it enters a latent state and integrates into the genome. Although lytic infections in cultured cells have been used to produce small amounts of recombinant proteins, it is the latent infection cycle which is of most interest because of the potential for long-term transgene expression *in vivo*.

AAV has many advantages as an *in vivo* expression vector, but there are also a number of problems that need to be overcome. Stable integration is very efficient in cells lacking a helper virus, and proviral transgenes can be expressed using either endogenous AAV promoters or heterologous promoters. However, if a transformed cell is subsequently infected with adenovirus or another helper, the AAV provirus is excised (rescued) and lytic infection commences. This capability must be eliminated from potential gene therapy vectors. The virus has a large host range that includes most dividing and post-mitotic cells. The full host range has yet to be determined but the success of the replication cycle depends on the host range of the helper virus. A strong advantage is that AAV is not pathogenic. No human diseases are associated with AAV infection, even though most people have been exposed to the virus, and up to 80% of the population has antibodies directed against AAV2 capsid proteins. Unfortunately, this is also a disadvantage, since the administration of AAV2 vectors tends to produce a humoral immune response. For this reason, some researchers have recently concentrated on the development of vectors based on other serotypes, particularly AAV5 and AAV6. Another advantage of AAV is safety: recombinant AAV vectors lacking replication functions require two helper viruses, a helper AAV to supply missing AAV functions, as well as adenovirus to supply functions missing in wild-type AAV. Finally, AAV shows no superinfection immunity, so cells can be transduced with several different AAV vectors to introduce multiple transgenes.

The integration of AAV into the host cell genome is more efficient in humans than other mammals, and this may reflect the specificity of the proviral insertion site. Most integration events involving wild-type AAV appear to occur within a relatively small region of chromosome 19 (19q13.1-qter), suggesting specific sequences may be involved that are endogenous to humans. This specific targeting could be an advantage in gene therapy, but recombinant AAV vectors tend to integrate randomly, indicating that the site specificity is conferred by AAV proteins. AAV has been shown to integrate into the DNA of all human cell lines tested, although the potential of primary cells has not been widely examined. In a recent study, recombinant AAV vectors have been used to achieve gene targeting by homologous recombination in human somatic cells (Chapter 6). One problem with AAV vectors is the low titer of recombinant viral stocks. During the initial development of AAV as a vector, this was as low as 10^4–10^5 transducing units per ml, although the careful optimization of preparation methods has increased titers routinely to 10^9.

AAV molecular biology

AAV2 binds to heparin sulfate proteoglycans on the cell surface, although other proteins may also be involved. The virion is internalized by receptor-mediated endocytosis and the single-stranded DNA is delivered to the nucleus. The AAV genome is approximately 5 kbp in length. The unique central region of the genome is flanked by 145 bp inverted terminal repeats, which are required for several functions including replication, gene regulation and proviral integration. The internal region of the genome is divided into two large sections (*Figure 4.8*). The *rep* region encodes non-structural proteins, and mutations within this region generate a replication-deficient phenotype that also lacks certain transcriptional

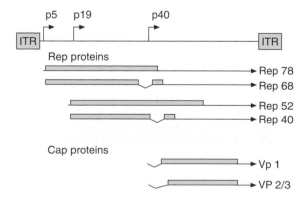

Figure 4.8

Organization of the AAV genome and its products. ITR = inverted terminal repeat.

functions. The *cap* region encodes the structural proteins of the capsid, and mutations within this region generate a packaging-deficient phenotype. There are three promoters in the AAV genome, named according to their map positions (p5, p19, p40). Rep proteins arise from mRNAs transcribed from the p5 and p19 promoters, while capsid proteins arise from mRNAs transcribed from the p40 promoter. All transcripts share a common intron and polyadenylation site. The polyadenylation site is adjacent to the origin of replication and *cis*-acting elements required for packaging, integration and rescue (*Figure 4.9*).

The regulation of AAV gene expression is complex, involving both helper virus and AAV functions. In the absence of helper virus, no AAV genes are expressed, but the genome may still undergo replication (catalyzed by host cell proteins). Depending on the cell type, the genome may integrate or remain as an episomal replicon, or it may be degraded. In the presence of adenovirus, transcription begins from the p5 and p19 AAV promoters, resulting in the synthesis of AAV Rep proteins. This in turn induces high-level transcription from all three AAV promoters, heralding the lytic cycle. The adenovirus E1a protein plays the major role in this process, as it is the initial transactivator of p5 and p19 transcription. However, further adenovirus gene products are also required for productive AAV infection (hence the HEK 293 cell line is not permissive for AAV replication). In fact, the E1b, E2a, E4 and VA genes are also required for the efficient accumulation and productive splicing of AAV mRNAs.

Construction of recombinant AAV vectors

The first AAV vectors, produced in 1984, were *cap*-replacement vectors, and transgenes were expressed at a low level from the endogenous AAV p40 promoter. Major improvements in AAV-mediated gene delivery came with the development of amplicon vectors containing just the polyadenylation site and the *cis*-acting sequences for packaging, integration and rescue in addition to the terminal repeats (*Figure 4.9B*). The maximum capacity of

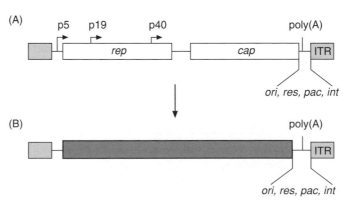

Figure 4.9

The *cap* and *rep* regions of the wild-type AAV genome (A) are completely replaced with foreign DNA in the highest-capacity AAV vectors (B).

the AAV capsid is 110% wild-type genome size, thus amplicon vectors of this nature allow the insertion of approximately 4.5 kbp of foreign DNA. The removal of the *rep* region provided an additional advantage: AAV Rep proteins control transcription as well as replication, and have been shown to interfere with endogenous promoters and enhancers in AAV vectors. However, *rep-* vectors are not affected in this manner and have been used successfully with several constitutive and inducible eukaryotic promoters.

One disadvantage of AAV expression vectors is the laborious procedure for producing stocks of recombinant virus. Since both a helper virus (adenovirus) and a helper AAV are required to provide missing functions in *trans*, there are two types of contaminant to remove. Wild-type AAV can be used as a helper, but the resulting recombinant stock contains adenovirus and wild-type AAV contaminants, the latter often in great excess to the recombinant virus. Much effort has been expended on developing procedures for the removal of contaminants. Adenovirus can be inactivated by heating (lysates are heated to 60°C for 2 hours) and effectively removed by CsCl density centrifugation or chromatography, but most laboratories now use helper plasmids to supply the missing adenoviral functions to avoid this laborious purification step. Several strategies have been employed to restrict the propagation of wild-type AAV, including the use of packaging-deficient AAV strains, the use of helper plasmids transfected into the cells infected with the parental AAV vector and adenovirus, the use of cell lines with integrated AAV genomes lacking functional rescue sequences, and the use of conditional lethal AAV mutants as helpers. These strategies have reduced the proportion of wild-type virus in the resulting stocks, although the level can still reach 10–50% due to recombination between the recombinant and helper AAV genomes. There is little chance of generating true AAV packaging lines because the overexpression of AAV genes is cytotoxic and, as discussed in the previous section, so is the overexpression of adenoviral helper functions. The most popular current method for producing essentially pure recombinant AAV is to use three plasmids, one carrying the recombinant AAV genome with all *cis*-acting sites intact but all *trans*-functions replaced by foreign DNA, one carrying the AAV rep and cap functions, but lacking any *cis*-acting elements, and the third supplying essential adenoviral functions. Commercially available systems include pXX2/pXX6, which provide the AAV and adenovirus functions on separate plasmids, and the pDG system, which provides both AAV and adenovirus helper functions on a single plasmid. There is no homology between the helper plasmids and transgene-containing plasmids, preventing the production of wild-type AAV by homologous recombination. However, with all the required functions supplied in *trans*, the recombinant AAV vector is rescued from the plasmid and packaged. (See references (9–14).)

4.4   Gene delivery with alphavirus vectors

Overview

The alphaviruses are single-stranded positive-sense RNA viruses that replicate in the cytoplasm. The wild-type genome has two genes, one encoding viral replicase and the other encoding a self-cleaving polyprotein containing all the capsid structural proteins. The genome acts as a direct substrate for translation, but initially only the 5' replicase gene can be translated because it possesses the typical cap structure required for eukaryotic protein synthesis. In contrast, the start of the 3' structural gene is internal and lacks such a cap. Naked viral RNA is infectious because the replicase gene is translated efficiently by host ribosomes. The replicase produces a full-length negative-sense antigenomic strand which acts as the template to generate full-length positive-sense daughter genomes. The same antigenomic strand also acts as the template to produce a subgenomic RNA carrying the structural proteins gene. Since this subgenomic RNA is also capped, the gene can be translated to yield structural proteins. The viral structural proteins are encoded on a separate subgenomic RNA to the replicase gene, so replication-competent vectors can be generated by replacing the structural protein region with foreign DNA.

Alphaviruses do not have a DNA stage in their replication cycle, so they cannot be used for stable transgene integration[1]. However, they are excellent transient expression vectors, reflecting their broad host range and cell tropism, their prodigious replication rate in infected cells (leading to high levels of recombinant protein) and the absence of cytotoxic effects. Alphaviruses are also very useful for short-term *in vivo* expression, as in cancer therapy. The two best-characterized alphaviruses are Semliki Forest virus (SFV) and Sindbis virus (SIN), and neither cause disease symptoms in humans. The delivery of recombinant viral RNA by transduction insures that neither the virus nor the transgene it carries integrates into the genome. One potential disadvantage with the use of alphaviruses and RNA viruses in general is the generally low fidelity of the viral replicase enzymes, which results in a higher level of transgene mutation than seen with DNA viruses.

Molecular biology of SFV and Sindbis virus

SFV, Sindbis and other alphaviruses infect cells by receptor-mediated endocytosis. The spike glycoprotein of the virus envelope then causes the viral and endosome vesicles to fuse, releasing the viral nucleocapsid into the cytoplasm. Proteins such as Sindbis spike glycoprotein have been exploited to enhance the efficiency of liposome-mediated transfection by catalyzing the fusion of liposomes and their target cells – such fusogenic DNA-containing particles are known as virosomes, and are discussed in more detail in Chapter 2. Following entry into the cytoplasm, the genomic RNA is released from the nucleocapsid and is immediately translated to yield the viral replicase (*Figure 4.10*). This enzyme produces progeny genomes by first synthesizing a negative-sense antigenomic strand, and then using this as a template to generate positive-sense genomic strands. The replicase also transcribes the subgenomic RNA from an internal promoter, and as discussed above, this encodes a polyprotein containing the viral structural

---

1 Alphavirus *transduction vectors* cannot be used for stable transformation. However, if a cDNA copy of an alphavirus genome placed under the control of a mammalian promoter is transferred to the cell by *transfection*, this can be stably integrated into the host genome and will yield large amounts of infectious viral genomic RNA.

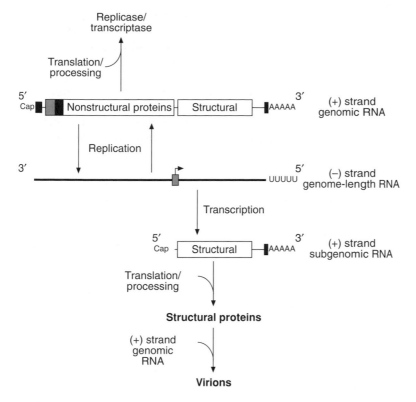

Figure 4.10

Alphavirus replication cycle. Translated regions of alphavirus genomic and subge-nomic RNAs are shown as boxes with the non-structural proteins and structural proteins indicated as open boxes. *Cis*-acting sequences important for replication and transcription are shown (grey boxes) as is the sequence in the non-structural region important for encapsidation (solid box). The start site for subgenomic mRNA transcription on the (−) strand genome-length RNA template is indicated by an arrow. Reprinted from Alphavirus based expression vectors: strategies and applications. Frarlov. I, *et al*, Pages 11371–11377. Copyright (1996), National Academy of Sciences, U.S.A.

proteins. The polyprotein is cleaved by capsid protein C, which has chymotrypsin-like autoproteolytic activity. The viral proteins associate with naked genomic RNA and the new nucleocapsids migrate to the cell surface where they are released by budding, generating new lipid envelopes with spike proteins.

Construction of recombinant SFV and Sindbis virus vectors

Because the viral structural proteins are encoded on a separate subgenomic RNA to the replicase gene, replication-competent vectors can be generated by replacing the structural protein region with foreign DNA. However, since the alphavirus replicase initiates transcription of the subgenomic RNA at an internal promoter, a further strategy for expressing recombinant proteins from alphavirus vectors is to add a second promoter either upstream or downstream of the structural genes, allowing the insertion of a foreign gene and its expression as an additional subgenomic RNA (*Figure 4.11A*). Alternatively, the insertion of an internal ribosome entry site (Chapter 7) between the structural genes and the transgene allows internal translation of the transgene, albeit at a lower efficiency compared to cap-dependent translation.

The production of replication- and assembly-competent vectors is not always desirable, particularly for *in vivo* applications, so the replacement of the structural genes with foreign DNA is often preferred. Such constructs are more efficient than the addition-type vectors, which tend to be unstable. Structural gene replacement does not affect replication, but it prevents the formation of infectious virions, and can result in yields of recombinant protein exceeding 50% of total cellular protein. Foreign DNA can be used to replace the entire structural coding region (*Figure 4.11B*), but the first 40 amino acid residues include a strong enhancer of protein synthesis, which significantly increases the yield of recombinant protein. In many expression systems, this region is included in the vector, so that the foreign gene is expressed as an N-terminal fusion protein (*Figure 4.11C*). Alternatively, the entire capsid C protein region can be included. In this case, foreign genes are initially expressed as N-terminal fusion proteins, but autocatalytic cleavage results in the production of native protein.

Since SFV and Sindbis are RNA viruses, *in vitro* manipulation and recombinant vector construction must involve the use of cDNA genome copies. These can be used to produce infectious RNA *in vitro*, which can be transfected into cells using many of the methods traditionally used for DNA transfection. For example, pSinRep5 is a Sindbis expression vector marketed by Invitrogen, which contains bacterial backbone elements, the Sindbis replicase genes and packaging site, and an expression cassette featuring a Sindbis subgenomic promoter, a multiple cloning site, and a polyadenylation site (*Figure 4.12*). There is an SP6 promoter upstream of the replicase genes and expression cassette for generating full-length in vitro transcripts. There is a second set of restriction sites downstream from the polylinker, allowing the vector to be linearized prior to *in vitro* transcription. Foreign DNA is cloned in the expression cassette, the vector is linearized and transcribed, and the infectious recombinant Sindbis RNA thus produced is

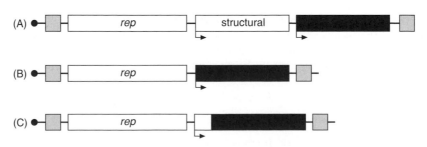

Figure 4.11

Structural gene replacement vectors based on alphaviruses. (A) Addition vector, (B) Standard replacement vector, (C) Standard replacement vector with endogenous translational enhancer.

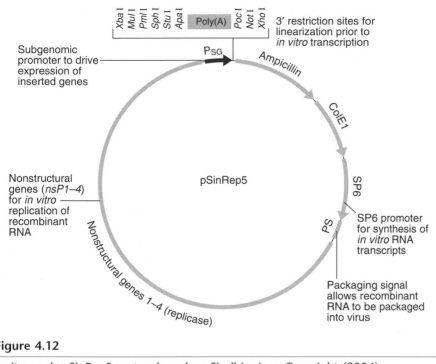

Figure 4.12

Invitrogen's pSinRep5 vector, based on Sindbis virus. Copyright (2004), Invitrogen Corporation. Used with permissions.

transfected into cells and expressed to generate high levels of recombinant protein. A different approach is to clone the entire alphavirus vector as an expression unit in a conventional plasmid expression vector under the control of a typical mammalian promoter, such as the SV40 promoter. In this case, DNA is transfected into the cell as normal and alphavirus RNA is produced as mRNA and exported to the cytoplasm. Here it replicates as a virus and produces large amounts of recombinant protein. In both the RNA and DNA transfection strategies, helper functions are not required and recombinant virus particles are not produced.

The transfection of recombinant RNA into cells is unsuitable for certain experiments, especially for gene delivery *in vivo*. In these cases, viral infection of cells is used for gene transfer. The propagation of infectious recombinant viruses requires helper functions, i.e. structural genes, to be supplied in *trans*. A binary approach has been used successfully to produce recombinant viruses. A vector such as pSinRep5 is used in concert with a second vector carrying the structural genes. Two *in vitro* transcription reactions are performed and target cells are cotransfected with two RNAs, one expressing replicase and the transgene, and one the structural proteins. This facilitates one round of replication and packaging and the production of recombinant viral particles that can be used to infect other cells. There is a risk of replication-competent viruses assembling by promiscuous replication, involving switching between the alternative RNA templates. This has been addressed by developing helper viruses with conditional lethal mutations. Another feature that limits the potential of alphavirus vectors for *in vivo* transfer is their tendency to shut down host cell protein synthesis and induce apoptosis. This could be advantageous for genetic vaccination, however, since it might help to elicit a strong immune response against the vector and the expressed transgene, and tolerance towards prolonged expression would not be a problem. (See references (15–20).)

4.5   Gene delivery with baculovirus vectors

Overview

The baculoviruses are a diverse group of double-stranded DNA viruses whose productive host range is limited to insects and other arthropods. There are no known vertebrate hosts, thus vectors derived from baculoviruses are among the safest to use in the laboratory. However, baculoviruses are taken up by mammalian cells in culture and recent studies have shown them capable of expressing foreign genes under the control of mammalian viral promoters. Baculoviruses are therefore potential vectors for gene therapy, and this relatively recent development is discussed in more detail below. However, first we consider the mainstream role of baculovirus vectors, which is transient expression of foreign proteins in insect cells. The usefulness of baculoviruses as vectors stems from the unusual infection cycle of one particular subfamily, known as the nuclear polyhedrosis viruses. The baculoviruses are divided into three subfamilies: the nuclear polyhedrosis viruses (NPVs), the granulosis viruses and the non-occluded viruses. The NPVs are potentially the most suitable vectors because they produce nuclear occlusion bodies, where mature virions are embedded in an abundant proteinaceous matrix. The matrix allows virions to survive in a harsh environment such as the external surface of leaves. The two important features of this system (as concerns vector development) are: (i) the matrix consists predominantly of a single virus protein, polyhedrin, which is expressed at very high levels; and (ii) the nuclear occlusion stage of the infection cycle is non-essential for viral propagation in insect cell lines. The polyhedrin coding region can therefore be replaced with foreign genes, allowing prodigious heterologous gene expression from the polyhedrin promoter, and such replacement vectors are replication-competent.

Because of the simple procedures involved in laboratory handling and propagation, vector development has concentrated on two species of virus. For the production of recombinant proteins in insect cells, the virus of choice is *Autographa californica* (alfalfa looper) multiple nuclear polyhedrosis virus (AcMNPV). This can be propagated in many insect cell lines, the most popular of which are derived from *Spodoptera frugiperda* (e.g. Sf9, Sf21). Alternative hosts include cell lines derived from *Estigmene acrea*, *Mamesta brassicae* and *Trichoplusia ni* (e.g. High Five™). A related virus (BmNPV), which infects the silkworm *Bombyx mori*, has been used for the production of recombinant protein in live silkworm larvae.

The baculovirus expression system has another important advantage in addition to its safety, convenience and yield. The rod-shaped viral capsid is completed after genome packaging, so that the size of the capsid is determined by the length of the genome. This means that any amount of foreign DNA can be accommodated, and multiple genes can be expressed in tandem. One disadvantage, at least for the production of mammalian proteins, is that there are certain differences in glycosylation between mammalian cells and many insect cells, although cell lines derived from *Estigmene acrea* produce glycan patterns that are more similar to mammalian glycans than those from *Spodoptera frugiperda*. Furthermore, mammalian glycosylation enzymes have been coexpressed with target foreign proteins using baculovirus vectors in Sf9 cells, and this strategy has successfully altered the specificity of the glycosylation pathway to generate correct modifications typical of mammalian proteins.

Molecular biology of AcMNPV

AcMNPV replicates with a biphasic cycle in susceptible insect cells, and produces two distinct forms of virus (*Figure 4.13*). The early cycle results in the production of extracellular budded viruses (EBVs), single enveloped virions that are released by budding from the cell membrane and go on to infect other cells. This first phase of the lytic cycle occurs within 20 hours following infection and involves the expression of three sets of viral genes. The immediate early (alpha) genes and the delayed early (beta) genes are expressed before DNA replication. The alpha genes are expressed immediately after the infecting virus is uncoated, and expression requires no viral gene products, hence transfected naked AcMNPV genomic DNA is infectious. The beta genes are expressed after the alpha genes and their expression is dependent upon alpha gene products. The late (gamma) genes

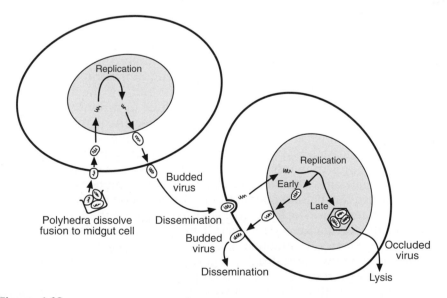

Figure 4.13

The baculovirus replication cycle in susceptible insect cells.

are expressed after the onset of replication and are thought to encode products involved in EBV structure and assembly. Most of the alpha, beta and gamma genes are essential for productive infection.

Twenty hours post-infection, the production of EBVs is dramatically reduced coincident with transcriptional depression of the alpha, beta and gamma genes, and a fourth set of very late (delta) genes is switched on. Partially assembled virions in the nucleus at this stage are enclosed within envelopes and then trapped in a paracrystalline matrix consisting mainly of a single protein, polyhedrin. The resulting structures are termed occlusion bodies, and function to protect the virus from temperature fluctuation and desiccation. The occlusion bodies are ingested by insects, and dissolve in the high pH of the digestive system, freeing the trapped viruses to infect the cells of their host. Several delta genes have been characterized, including the polyhedrin gene itself, the p10 gene whose function is unknown, and several genes encoding components of the occlusion body envelope. Importantly, these genes can be regarded as non-essential for productive infection in insect cell lines, although they are essential for the spread of the virus in nature. This establishes a containment system for laboratory-constructed baculovirus vectors.

Construction of recombinant baculovirus vectors

Most baculovirus vectors are replacement vectors in which either the polyhedrin or p10 coding regions are substituted with foreign DNA. As discussed above, the non-essential nature of these delta gene functions has made them desirable targets for replacement with foreign DNA. There are also vectors which use early promoters, particularly for the production of proteins known to be toxic to insect cells, and more recently for baculovirus surface display technology. Polyhedrin replacement vectors are most popular due to the prodigious expression of polyhedrin in the late part of the replication cycle (accounting for up to 25% of total cellular protein, or 1 mg per 10^6 cells) and the ease with which recombinant vectors can be identified (see below). The polyhedrin upstream promoter and 5′ untranslated region are important for high-level foreign gene expression and these are included in all polyhedrin replacement vectors. Initially, the highest levels of recombinant protein expression were obtained as fusions with at least the first 30 amino acids from the N-terminal region of the polyhedrin protein, and although it was initially thought that the N-terminal region increased protein stability, it was later shown that additional regulatory elements were present in this region, located downstream of the polyhedrin translation start site. For the production of native proteins, vectors are available where the natural polyhedrin initiation codon is mutated, so that these important 'downstream' sequences become part of the 5′ untranslated region of the foreign gene. However, it has been reported that initiation may still occur at this mutated site, so cloned foreign genes must be trimmed of their own untranslated regions, and the start codon should be out of frame with respect to the natural polyhedrin start codon.

The original method for identifying recombinant polyhedrin-replacement clones was screening plaques for lack of occlusion bodies. The wild-type virus produces occlusion bodies, which cause plaques to appear opalescent under an oblique light source. Conversely, recombinant plaques lack occlusion bodies (OB–) and therefore appear clear. Such analysis must be carried out using a light microscope as baculovirus plaques are very small. The major disadvantage of p10 replacement vectors is that p10 mutants are not OB– and have no easily scorable phenotype. Plaques of both polyhedrin and p10 replacement vectors can be screened for the presence of the insert by hybridization or immunological detection of foreign protein. More recently, a number of powerful visual screening strategies have been developed as well as systems for selecting recombinant viruses. The *E. coli lacZ* gene has been used to help identify recombinant plaques. In the simplest strategy, the general visibility of plaques is improved by insertion of the *lacZ* gene under an appropriate promoter somewhere in the baculovirus genome, so that all plaques turn blue upon exposure to X-gal. More refined approaches include exploiting *lacZ* for blue-white selection: by using parental baculovirus strains in which β-galactosidase is expressed from the polyhedrin promoter, recombinants (which replace the *lacZ* gene with the foreign gene to be expressed) form clear plaques, while parental vectors form blue plaques; alternatively, by introducing *lacZ* as a marker alongside the foreign gene, the recombinant vectors form blue plaques while the wild-type virus produces clear plaques. Due to the lack of screenable phenotype, p10 expression vectors must incorporate a reporter gene detection system to allow recombinant plaques to be identified. Recently, vectors have been designed with polyhedrin expressed from the p10 promoter, so that the original OB assay can be used.

Baculovirus genomes are large, and although strains have been constructed with unique restriction sites, allowing insertion of foreign DNA by in vitro ligation, the favored strategy is homologous recombination using plasmid targeting vectors containing a baculovirus homology region into which foreign DNA has been inserted. Generally, plasmid and wild-type baculovirus DNAs are cotransfected into the appropriate insect cells either by calcium phosphate transfection, lipofection or electroporation. This strategy generates recombinant vectors at a frequency of 0.5–5%. Linearized baculovirus genomes are non-infectious, but remain recombinogenic, and this can be used to reduce contamination from wild-type virus. The proportion of recombinants can be vastly increased through the use of non-viable deleted derivatives of the wild-type baculovirus genome, which are repaired by homologous recombination with the targeting vector. Derivatives of the wild-type AcMNPV genome, with unique restriction sites added upstream of the polyhedrin gene and within an essential gene found downstream of the polyhedrin locus, can be used to generate linear genome fragments lacking an essential function. Such non-viable linear genomes are now commercially available (e.g. BacPAK6). Compatible targeting vectors span the deletion and provide enough flanking homologous DNA to sponsor recombination between the two elements and generate a viable, recombinant genome. Such approaches result in the production of up to 90% of recombinant plaques. Combinatorial approaches using deleted non-viable genomes and targeting vectors incorporating *lacZ* visible screening systems provide very powerful selection for recombinant vectors. Even the minor inconvenience of waiting one day for the blue-white screening assay to develop has been overcome with the introduction of baculovirus vectors using immediately visible markers such as green fluorescent protein.

Alternative systems, in which the baculovirus genome is maintained and targeted as a low copy number episome in bacteria or yeast, are gaining popularity because they allow the direct isolation of recombinant vectors. Low copy number maintenance is important to prevent the survival of a background of non-recombinant vectors. The baculovirus genome can be stably maintained as a low copy number episome in bacteria if it contains an F-plasmid origin of replication and a bacterial selectable marker such as kanamycin resistance. This system, marketed by Gibco-BRL under the name 'Bac-to-Bac', exploits the specificity of the bacterial transposon Tn7 to introduce foreign genes into the baculovirus/plasmid hybrid, which is called a bacmid. The foreign gene is cloned into a bacterial transfer plasmid between two Tn7 repeats. This is introduced into the appropriate strain of *E. coli*, which contains the bacmid and a helper plasmid supplying Tn7 transposase. Induction of transposase synthesis results in the site-specific transposition of the transgene into the bacmid, generating a recombinant bacmid that can be cloned and isolated from bacterial culture for transfection into insect cells. The Tn7 target site in the bacmid is inserted in-frame within the *lacZ* gene, allowing blue-white screening of recombinants, and rapid isolation of pure recombinant bacmid DNA from bacterial culture.

The baculovirus genome can also be maintained as a low copy number episome in yeast if it contains a suitable origin of replication, a centromere and selectable markers: these elements are inserted as a cassette to replace the polyhedrin gene. In the original system, a pair of selectable markers was used, allowing the power of yeast genetics to be applied to vector selection. One marker was used for positive selection of transformed cells, while the other was used for counterselection against non-recombinant vectors. The *SUP4-o* marker was initially used for counterselection. This is a nonsense suppressor that, in the particular yeast strain used, confers sensitivity to the arginine analog canavanine and the ability to grow in media lacking adenine. Removal of the marker by replacement with homologous DNA confers resistance to canavanine and a requirement for adenine. Plasmid DNA isolated from canavanine-resistant, adenine-requiring yeast cultures was used to transfect insect cells and produce pure recombinant baculovirus. In both bacteria and yeast, the maintenance sequences (origin of replication/centromere and positive selection marker) must stay in the recombinant baculovirus vector. They have been shown to have no effect on baculovirus replication or gene expression in insect cells.

Baculovirus-mediated gene transfer to mammalian cells

The ability of baculoviruses to infect mammalian cells was first demonstrated in 1967. However, it was not until the mid 1990s that recombinant baculovirus vectors containing genes under the control of mammalian promoters were first used to transduce and express foreign proteins in mammalian cells. In the first reports, the human cytomegalovirus (CMV) promoter was used to drive luciferase gene expression and the Rous sarcoma virus long terminal repeat promoter (RSV-LTR) was used to drive *lacZ* gene expression. A variety of cell lines was tested and it appeared that efficient gene expression was possible only in cells of hepatic origin. Since then, however, many different mammalian promoters have been used and the range of amenable cell lines has been vastly extended, including primary cell lines and post-mitotic cells.

While representing an efficient *in vitro* transduction system, the inability of baculoviruses to replicate in mammalian cells makes them especially valuable as a safe alternative for *in vivo* applications such as DNA vaccination and gene therapy. Unfortunately, early studies of *in vivo* gene transfer using baculoviruses were hampered by the sensitivity of the virus to complement-mediated inactivation. The introduction of baculovirus vectors into mice in which the complement system had been inactivated resulted in efficient gene transfer and expression. Therefore, a baculovirus vector has been developed expressing its own regulator of complement activity, decay-accelerating factor, and has been shown to be highly efficient for *in vivo* gene delivery and expression. Efficient gene transfer has also been achieved by pseudotyping the baculovirus with a vesicular stomatitis virus G glycoprotein. Transfer by unmodified baculovirus vectors is efficient if the virus is not exposed to high levels of complement, as occurs for example in the brain.

Baculoviruses are useful not only for the delivery of foreign genes into mammalian cells, but also for the delivery of other viruses. For example, hepatitis C virus does not infect cultured cells, but a hybrid baculovirus containing the entire HCV genome can initiate an HCV infection. Baculoviruses can also be used to improve the production of recombinant viral vectors. As discussed above, vectors based on adenoviruses are generally replication-defective because they lack one or more essential viral gene products, and the highest-capacity amplicons or fully deleted adenoviruses (FD-AdV), contain no viral genes at all, only those *cis*-acting elements required for replication and packaging. Cell lines are not available for the packaging of such vectors so helper viruses are normally required, which leads to contamination of recombinant stocks. Recently, however, a recombinant baculovirus vector has been developed which carries a packaging-deficient copy of the entire adenovirus genome. This hybrid is maintained stably in insect cells and supplies all the required adenoviral functions when introduced into HEK 293 cells, allowing the adenovirus amplicons to be packaged efficiently. (See references (21–27).)

4.6   Gene transfer with herpesvirus vectors

Overview

The herpesviruses are large, enveloped viruses with linear, double-stranded DNA genomes varying in length from 100-200 kbp. The virion structure is complex, consisting of an outer lipid envelope studded with a large number of glycoproteins and other proteins, surrounding a matrix of proteins known as the tegument, which overlies the capsid proper. The complexity of the lipid envelope means that the herpesviruses demonstrate considerable variability in their host range and cell tropism. Furthermore, the same virus may promote the lytic infection of some cell types and the latent infection of others, the latter resulting in long-term episomal maintenance of the viral genome. The two herpesviruses that have been developed as vectors are Epstein-Barr virus (EBV) and herpes simplex virus (HSV1). EBV has a very narrow host range and tropism (limited mainly to human B-lymphocytes and a few other cell types), so it is not much use as a general transduction vector. However, the EBV genome will replicate in many cells following transfection, wherein it is maintained as an episomal replicon. The EBV genome has therefore been used as a basis of a series of large-capacity plasmid vectors for episomal transgene expression in human cells (see Chapter 2). In contrast, HSV1 has a very wide host range and cell tropism, and has been developed into a versatile transduction vector. Following infection, the HSV genome remains within the nucleus of the host cell indefinitely, i.e. it has a life-long latent infection cycle. The lytic cycle can be reactivated by a variety of stresses in cells infected with the wild type virus, but HSV vectors, with the lytic functions deleted, can be maintained permanently in a latent state, promoting long-term transgene expression. Other advantages of HSV1 as a vector include its large capacity for foreign DNA, its efficient infection of neurons, and the ability of replication-competent viruses to cross synapses, a phenomenon that has been exploited for tracing neuronal pathways.

Molecular biology of HSV-1

HSV-1 has a broad host range and cell tropism because glycoproteins C and B, embedded in the envelope, interact with ubiquitous glucosaminoglycans on the cell surface, e.g. heparan sulfate and dermatan sulfate. Following this initial binding, another envelope component – glycoprotein D – then interacts with specific protein receptors, known as herpesvirus entry mediators (e.g. HveA, a member of the tumor necrosis factor-α/nerve growth factor receptor family). These interactions result in the fusion of the viral envelope and the cell membrane, and the release of the nucleocapsid into the cytosol. From the nucleocapsid, the linear DNA is transferred into the nucleus whereupon it immediately circularizes. The capsid also contains a virion host shut-off protein (VHS), which interrupts host protein synthesis, and a transcriptional activator termed VP16. During lytic infection, VP16 forms a dimmer with the host transcription factor Oct-1 and induces the expression of the viral immediate early genes ICP0, ICP4, ICP22, and ICP27. The immediate early genes are all essential for lytic replication. These encode further regulators, which act both on their own genes (in a self-regulatory manner) and upon a set of approximately 15 early genes controlling DNA replication. Following replication, approximately 40 late genes are activated, which encode DNA cleavage and packaging proteins and capsid proteins. Progeny genomes are then packaged into virons and transported to the cell surface, where lysis occurs (*Figure 4.14*).

The switch to latency is thought to reflect the balance of host-and virus-encoded transcriptional regulators in the cell. The immediate early genes are not expressed, and viral activity is restricted to a 152 kbp genomic region that overlaps the immediate early gene ICP0 in the antisense orientation. This region encodes a set of latency-associated transcripts (LATs). The LATs are also synthesised during lytic infection, although no protein products have been detected (even though LATs are associated with

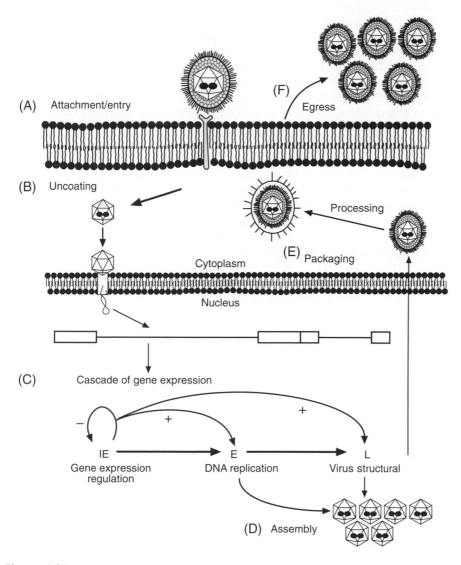

(A) Attachment/entry

(F) Egress

(B) Uncoating

Processing

Cytoplasm

(E) Packaging

Nucleus

(C) Cascade of gene expression

−   +   +

IE ⟶ E ⟶ L

Gene expression   DNA replication   Virus structural
regulation

(D) Assembly

Figure 4.14

The replication cycle of herpes simplex virus. Copyright (2004), Humana Press. Used with permissions.

ribosomes). During latency, the LATs remain in the nucleus. The LATs are neither required for the establishment of latency, nor its maintenance, but they are required for reactivation of the lytic cycle.

Strategies for HSV vector construction.

Three different types of HSV1 vector have been developed for gene transfer. Replication-competent viruses can be generated by inserting foreign DNA between viral genes or replacing non-essential (accessory genes) by homologous recombination, usually by transfecting HSV-infected cells with a targeting vector containing a foreign gene within a viral homology region. Such vectors can be propagated in culture without the use of a complementary cell line, but the removal of non-essential genes may nevertheless abolish functions, which contribute to pathogenicity in natural infections, thus rendering the vectors safer than the wild type virus. The problem with such vectors is that they can provoke cytotoxic responses and are generally immunogenic.

For most gene therapy applications, it is better to use replication-defective vectors. Several derivatives of HSV are now available that carry deletions in one or more of the immediate early genes, and such vectors must be propagated in a complimenting cell line or in the presence of a cotransfected helper plasmid. The deletion of immediate early genes ICP4, ICP22, ICP27 and ICP0 is also beneficial because their products are toxic to the host cell. However, the deletion of ICP0 also appears to reduce transgene expression levels in infected cells. The major limitation of these vectors is the inability to achieve prolonged transgene expression.

The final category of HSV vector is the amplicon, a plasmid based mini-HSV vector containing the HSV origin of replication, cleavage and packaging sites, but no viral genes. Such vectors can only be packaged in the presence of a helper virus supplying the many missing functions in trans. Wild type HSV-1 can be used to supply helper functions, but the wild type virus often causes lytic infections, resulting in rapid cell death both *in vitro* and in live animals. A number of HSV-1 mutants have been developed as helper viruses, as these are nonpermissive for lytic replication (but still induce latent infection when introduced into cultured neurons or injected into the brain). Initially, a temperature sensitive mutant was used carrying a point mutation in one of the immediate early genes. This was conditionally defective, inducing latent infections in the brain and in cultured neurons at 37°C but lytic infections at 31°C. Unfortunately, its applicability was limited by a significant reversion frequency, resulting in the induction of lytic infections. More recently, defective helper viruses with deletions in one or more of the immediate early genes have been used, providing complementary packaging lines to produce infective recombinant particles. In some cases, the reversion frequency has been reduced to 10^{-7}.

HSV vectors for prolonged transgene expression in neurons.

Because of their ability to achieve long-term latent infection of neurons, the major use of HSV-1-derived vectors has been for gene transfer to neurons either *in vitro*, or in the central nervous system of living animals. HSV-1 vectors possess a number of advantages for gene transfer to neurons, including efficient DNA transfer, a large genome size permitting the transduction of large segments of foreign DNA, and long term episomal maintenance. Unfortunately, transgene expression usually falls off within a few days of infection, a phenomenon which appears to be unlinked to the persistence of the vector itself, and also independent of the promoter used to drive the transgene. A wide variety of promoters has been used, including HSV immediate early gene promoters, constitutive viral promoters typically used in plasmid vectors (e.g. SV40, human cytomegalovirus) and even neuron-specific promoters from mammals. To address this problem, several researchers have attempted to use the transcriptional control elements that regulate the expression of LATs during latent infection, two of which have been identified – LAP1 and LAP2, and continued transcription has been detected for several weeks after inoculation.

Modifying the tropism of HSV

One aspect of HSV vector development which has been widely studied is the modification of cellular tropism, which may improve the versatility of the virus as a gene therapy vector. Several different approaches have been considered, including modification of the lipid envelope proteins so the virus is absorbed to a defined set of cell types, modification of the structure of glycoprotein D to dictate the receptors with which it can interact, the use of soluble receptors and adapters, and pseudotyping the HSV envelope with envelope components from other viruses. As an example of the above, HSV mutants has been generated in which glycoprotein C was fused to part of the hormone erythropoietin. This reduced the affinity of the virus for heparan sulfate and promoted interactions with cells expressing the erythropoietin receptor. (See references (28–33).)

4.7 Gene delivery with retrovirus vectors

Overview

Retroviruses are enveloped RNA viruses, which comprise a proteinacious virion approximately 100 nm in diameter surrounded by a lipid bilayer that facilitates fusion with the plasma membrane of the host cell. Each viral particle carries two copies of a single-stranded, positive-sense RNA genome as well as several proteins required for infection. There are seven genera of retroviruses based on data from sequence comparisons, but they are often described as either simple or complex, the former referring to the conventional oncoretroviruses, such as murine leukemia virus (MLV) and the latter to the lentiviruses, such as human immunodeficiency virus (HIV) and spumaviruses which contain additional genes compared to the basic oncoretrovirus genome (see below). The replication strategy of the retroviruses is unique. After entering the cell, the virus is uncoated and the genomic RNA is transported to the nucleus where it is converted into a terminally redundant double-stranded cDNA copy by the virion protein reverse transcriptase. A second virion protein, integrase, then inserts this cDNA copy into the host genome. The integrated provirus is transcribed to yield progeny RNA genomes as well as subgenomic mRNAs encoding enzymes and structural proteins of the viral capsid. The host range of each retrovirus species is determined by proteins in the lipid envelope. Some retroviruses have a very restricted host range (e.g. HIV, whose primary receptor is the immune recognition protein CD4 found only on T-cells and macrophages) while others have a broad host range because the envelope proteins interact with more widely distributed receptors. The envelope proteins of MLV are particularly promiscuous, so this virus has a broad host range and cell tropism and has been extensively developed as a vector for gene transfer to mammalian cells.

The retroviruses have long been known to be capable of transducing exogenous genes into host cells. Indeed the discovery of the oncoretroviruses reflected their ability to cause tumors in infected animals through the integration and expression of oncogenes derived from cellular genes responsible for the control of cell growth and proliferation. Through the analysis of such acute transforming retroviruses, over 100 oncogenes and their cellular counterparts have been discovered and characterized. In many cases, the oncogene is expressed as a fusion protein with one of the viral gene products, but in some viruses the oncogene is expressed as an independent transcription unit. These latter examples, which include for example the v-*src* gene in Rous sarcoma virus, provide an immediate strategy for vector development and transgene expression. There are even natural examples of oncoretroviruses carrying two unrelated oncogenes (e.g. v-*erbA* and v-*erbB* in avian erythroblastosis virus).

Retroviruses are advantageous vectors for numerous reasons, not least of which are the ability to produce high viral titers (10^6–10^8 particles ml^{-1}) using readily available packaging lines, the incredible efficiency of stable transduction (approaching 100% *in vitro*, but also very high *in vivo*) and the ability to pseudotype viral particles and thus engineer the host range of each vector. Furthermore, the small viral genome is easy to manipulate in the laboratory once it has been converted into a cDNA copy, and it carries a useful promoter/enhancer system, which can be used to drive transgene expression. In some viruses, this regulatory system is constitutive but in murine mammary tumor virus (MMTV) it is inducible by glucocorticoid hormones and synthetic analogs such as dexamethasone, providing a natural system for inducible transgene expression (see Chapter 7). Because of the high efficiency of transduction and the facility for stable transgene integration, retroviruses were among the first viral vectors to be investigated for gene therapy. Indeed, the first successful gene therapy trial (for the treatment of severe combined immunodeficiency (SCID) caused by adenosine deaminase (ADA) deficiency) was carried out using a retroviral vector (page 241). The efficiency of retroviral transduction *in vivo* means that retroviruses are also used to introduce marker genes in developing mammalian and avian embryos for lineage-tracing studies and the expression or inhibition of developmental regulators. One area in which this approach has been particularly useful is the study of limb development, where Rous sarcoma virus (RSV) vectors have been widely used to transduce developing chicken embryos. Also, as discussed in Chapter 5, retroviruses are capable of infecting eggs and early embryos, and have therefore been extensively used to generate transgenic animals and birds. An important difference between oncoretroviruses and lentiviruses in terms of their use as vectors is that the oncoretroviruses *can only infect dividing cells* whereas lentiviruses are not similarly constrained.

Molecular biology of the retroviruses

The genome organization of the oncoretroviruses is simple and highly conserved (*Figure 4.15A*). The integrated provirus comprises three major open reading frames (*gag*, *pol* and *env*) bracketed by so-called long terminal repeats (LTRs). A single promoter located in the left LTR is used to transcribe genomic RNA. The *gag* region encodes viral structural proteins (group antigen), the *pol* region encodes the reverse transcriptase and integrase required for genomic integration in the host cell as well as a protease required to process viral gene products, and the *env* region encodes viral envelope proteins. The full-length RNA is translated to produce the Gag and Gag-Pol polypeptides which are processed by the protease, and a splicing event produces subgenomic RNA that is translated to generate the Env polypeptide. Lentiviruses have two additional groups of genes that have essential functions during the viral life cycle and pathogenesis (*Figure 4.15B*). These are the regulatory genes (*Tat* and *Rev*) and the accessory genes (*Vpu, Vpr, Vif* and *Nef*). The Tat protein regulates the promoter activity of the 5′ LTR promoter and is necessary for transcription of the viral genome. The Rev protein binds to a Rev response element (RRE) within the viral RNA, allowing the transport of unspliced RNA out of the nucleus, and is necessary for efficient *gag* and *pol* expression. The accessory genes are essential virulence factors *in vivo* but are dispensable *in vitro*.

The retroviral infection cycle begins with the fusion of the viral envelope and host cell plasma membrane, brought about through an interaction between virus envelope proteins and the appropriate cell surface receptors (*Figure 4.16*). The packaged RNA genome is shorter than the integrated provirus and does not have long terminal repeats. This is because it is transcribed from an internal promoter in the left proviral LTR and finishes at a polyadenylation site located within the right proviral LTR. The LTRs are regenerated during the replication cycle in which the RNA genome is converted into a cDNA copy. The cDNA then integrates into the genome of the host cell in a manner similar to that used by DNA transposons, and is immediately transcribed to yield mRNAs and progeny genomes. A *cis*-acting site termed ψ, which is located at the 5′ end of the genome, is required for packaging the genome into new virions.

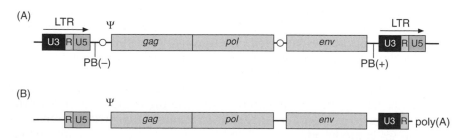

Figure 4.15

Structure of a generic oncoretroviral genome (A) before and (B) after integration. Upper figure shows the structure of an integrated provirus, with long terminal repeats (LTRs) comprising three regions U3, R and U5, enclosing the three open reading frames *gag, pol* and *env*. Lower figure shows the structure of a packaged RNA genome, which lacks the LTR structure and possesses a poly(A) tail. PB represents primer-binding sites in the viral replication cycle, and ψ is the packaging signal. The small circles represent splice sites.

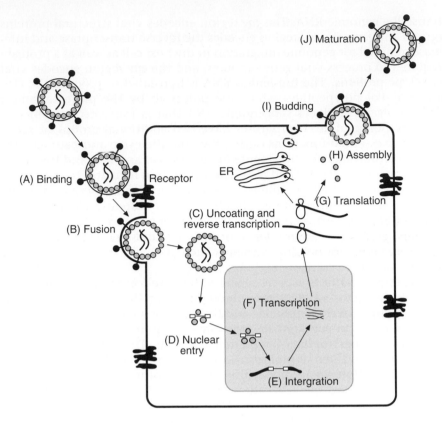

Figure 4.16

The retroviral life cycle. Copyright (2004), Humana Press. Used with permissions.

Construction of recombinant oncoretroviral vectors

Although replication-competent oncoretroviral vectors have been developed in which the transgene is inserted in addition to the essential viral genes, such vectors have a limited capacity and replication-defective replacement vectors are preferred. These usually are fully deleted derivatives, lacking the *gag, pol* and *env* open reading frames but retaining the essential *cis*-acting sites including the LTRs, primer binding sites for genome replication, the packaging site ψ and, if necessary, the splice signals (*Figure 4.17*). Importantly, the ψ site extends into the *gag* coding region and the entire sequence (ψ+) must be included to insure a high titer of recombinant viruses is produced.

Such replication-defective vectors can be propagated only in the presence of helper viruses or in a suitable packaging cell line, which generates viral particles competent for infection, reverse transcription and genome integration, but prevents the propagation of recombinant viruses once the intended host cell has been transduced. Various packaging lines have been developed allowing different forms of pseudotyping. Vectors with a broad host range are generated using packaging lines derived from amphotropic viruses such as MLV but many alternative lines are available, which allow

Figure 4.17

Replication-defective retroviral amplicon vectors. (A) The transgene has replaced the *gag*, *pol* and *env* genes, and is driven by the endogenous retroviral LTR promoter in U5. (B) The LTR promoter has been mutated (x), and the transgene is driven by its own promoter.

the tailoring of vector host range for particular experimental strategies. Packaging lines also vary in terms of the configuration of the viral helper functions, and this helps to prevent the adventitious production of replication-competent viruses by recombination. The most efficient lines contain helper viruses with genomes modified to limit the extent of homologous sequence shared between the helper virus and the vector, and increase the number of independent crossover events required to form a replication-competent genome. One of the most efficient lines in this category is GP + E-86 which contains split coding regions, point mutations and deletions, and hence requires three independent recombination events to generate a replication-competent virus (*Protocol 4.2*).

Many different strategies have been developed to express transgenes in retroviral vectors. The most straightforward is to remove all the viral coding sequences and place the transgene between the LTR promoter and the viral polyadenylation site. This is suitable where constitutive transgene expression is acceptable, but if the transgene needs to be controlled in a more sophisticated manner, then the LTR promoter must be inactivated and regulatory elements for transgene expression must be included in the construct. Inactivation of the LTR promoter is often achieved through the use of so-called suicide vectors (also known as self-inactivating vectors) in which a deletion is included in the 3′ LTR of the packaged genome. During the next round of genome replication, this deletion is copied to the 5′ LTR of the provirus, destroying their promoter activity. Any internal promoters remain unaffected.

As in transfection experiments, it is often necessary to use selectable markers to identify stably transduced cells. This requires two transgenes – the marker and the primary transgene – to be expressed from the same vector. There are several ways to achieve this (*Figure 4.18*). In one strategy, two independent transcription units can be constructed within the virus, so that the two transgenes are controlled by separate promoters. Alternatively, the vector can be modeled on the splicing pattern of a wild-type virus, so that full-length and spliced transcripts are produced. In both cases, it is essential to allow full-length RNAs to be produced from the recombinant genome in the packaging line, so it is important *not* to include a polyadenylation site at the end of the first gene. In two-promoter vectors, the lack of a polyadenylation site for the upstream gene can result in read-through transcription and occlusion of the second promoter. This can be

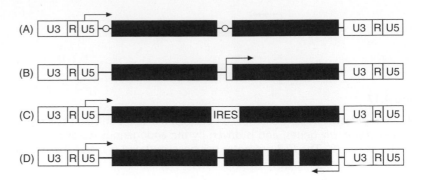

Figure 4.18

Strategies for dual transgene expression in retroviral vectors. (A) Both genes driven by LTR promoter after splicing. (B) Proximal gene driven by LTR promoter, distal gene by its own promoter. (C) Both genes transcribed from the same LTR promoter, distal gene translated separately using internal ribosome entry site. (D) Dual transgenes arranged in opposite orientations, proximal transgene driven by LTR promoter, distal transgene contains introns and is driven by an opposing promoter.

alleviated by placing the two genes in opposite orientations, so that the (reversed) polyadenylation site of the transgene is not recognized during full-genome transcription. For alternative splicing vectors, the possible existence of cryptic splice sites the upstream gene should be considered. Multiple gene expression with LTR-based, internal promoter-based and alternative splicing vectors can also be achieved by using internal ribosome entry sites (page 185). As well as polyadenylation sites, the presence of introns in the transgene may also interfere with viral replication and these should be avoided wherever possible. If the inclusion of an intron is unavoidable, the transgene should be placed in the reverse orientation under the control of an independent promoter, so that the introns are not recognized and spliced during genomic transcription in the forward orientation.

Special considerations for the construction of lentivirus vectors

Lentiviral vectors are produced in much the same way as MLV and other retroviral vectors, i.e. by replacing essential viral genes with the transgene of interest and using a packaging line to supply the missing viral functions. Most interest has been shown in vectors based on HIV, but systems have also been described that are based on bovine immunodeficiency virus (BIV), equine infectious anemia virus (EIAV), feline immunodeficiency virus (FIV) and simian immunodeficiency virus (SIV). Special considerations for such vectors include the requirement for Tat and Rev functions, and the fact that HIV is one of the few retroviruses known to cause an infectious disease in humans. Therefore, there must be especially stringent precautions to prevent contamination with replication-competent virus when HIV is used as a vector, and this has prompted the development of a series of multicomponent packaging lines in which different viral functions are supplied on different plasmids introduced by transient transfection.

The first generation of lentiviral packaging lines involved transfection of 293T cells with three plasmids, one containing a wild-type HIV genome with the *gag, pol* and *env* regions replaced by a transgene, one providing the HIV *gag* and *pol* functions under the control of a constitutive promoter and one encoding the G-protein from vesicular stomatitis virus (VSV) to replace the HIV *env* function. The combination of these three plasmids resulted in the production of VSV-G pseudotyped lentiviral vectors at a high titer, but some pseudotyped replication-competent HIV were produced by recombination during viral replication and packaging. In the second generation of vectors, the *vpr, vpu, nef* and *vif* genes were removed from the first plasmid, since these genes are important for virulence *in vivo* but are not required for productive infection in cell lines. More recently, a third generation of vectors has been produced in which the U3 region of the 5′ UTR has been replaced by the human cytomegalovirus early promoter, therefore making the activity of the LTR independent from the *Tat* gene, which consequently has been deleted. The U3 region of the 3′ UTR has been modified to include a deletion, making the vector self-inactivating, and several internal modifications have been made including the incorporation of sequences that enhance transgene expression (*Figure 4.19*). The *Rev* gene, which is required for *gag* and *pol* expression, has also been removed from the packaging construct and is instead supplied from a fourth plasmid to further reduce the likelihood of replication-competent vectors arising by recombination. Indeed the presence of four plasmids sharing no significant homology, makes such recombination events very unlikely even at high production titers. (See references (34–42).)

4.8   Gene delivery with poxvirus vectors

Overview

The poxviruses are DNA viruses with large genomes (up to 300 kb) and a wide host range, including many vertebrate and invertebrate species. Unusually for DNA viruses, replication takes place in the cytoplasm of the host cell, and therefore the virus cannot exploit any of the host's replication or transcription machinery. All the enzymes and accessory proteins required for these processes must be encoded by the virus. As vectors, the potential of poxviruses has been exemplified by vaccinia virus, first used to eradicate smallpox and then developed as a recombinant vaccine carrying epitopes from other infectious organisms. Although much attention has focused on the role of vaccinia and other poxviruses in immunization, vaccinia virus and to a lesser extent other poxviruses such as canarypox represent useful transient expression vectors for a range of cell lines, reflecting their wide host range and capacity for high-level transgene expression.

Molecular biology of vaccinia virus

Vaccinia virus has a complex proteinacious core surrounded by a lipid envelope derived from the Golgi apparatus of the infected cell. Enveloped

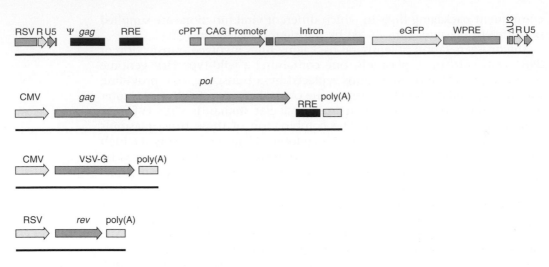

Figure 4.19

Prototypical plasmids for the production of 3rd generation HIV-1 lentiviral vectors. In the top panel, the U3 region in the 5'LTR is replaced by an RSV promoter and a deletion has been introduced in the U3 region of the 3'LTR (DU3). The ψ region and a small part of the *gag* coding sequence necessary for packaging, the rev response element (RRE), and the central polypurine tract (cPPT) are the only regions left from HIV-1. The transgene, in this case *GFP*, is driven by a CMV/β-actin promoter (CAG). For mRNA stabilization, a Woodchuck post-transcriptional response element is included (WPRE). The lower three panels show helper plasmids for the production of lentiviral vectors. The first plasmid contains only the coding regions for the HIV *gag* and *pol* genes and the Rev response element (RRE) necessary for efficient splicing. The coding region for the accessory genes, regulatory genes, and envelope, the packaging signal (ψ), and the LTRs are removed. The second plasmid codes for the VSV-G envelope. The *rev* gene is expressed on the third plasmid. Reprinted from Methods, Vol 33, No 2, Blesch, A. "*Lentiviral and . . .,*" pages 164–172, Copyright (2004), with permission from Elsevier.

virions are termed extracellular enveloped viruses, but the majority of infectious particles are found as intracellular naked viruses in the cytoplasm. The genome is double-stranded linear DNA, and the ends of the DNA strands are sealed by hairpins. There are four functional sets of genes, knows as the constant, immediate, intermediate and late genes. The immediate genes are required for genome replication and uncoating. They are expressed at the beginning of the infection and have early promoters activated by the viral RNA polymerase carried in the virion. After replication, the immediate genes are switched off and the intermediate genes are switched on. Some of these encode transcriptional regulators that activate the late genes, which encode components of the capsid and packaging proteins. The constant genes have both early and late promoters and are hence expressed throughout the infection cycle.

Construction of recombinant vaccinia virus expression vectors

Vaccinia virus is simple to grow because of its broad host range, including both established cell lines and primary cells. However, the efficiency of plating varies according to cell type. The large genome size and unusual replication strategy represent major obstacles to the design and construction of expression vectors. The genome is too large to manipulate in vitro, and because the virus normally carries its own replication and transcription machinery into the cell, recombinant genomes introduced into the cell by transfection are not infectious. Although it is now possible to generate infectious recombinant genomes, the strategy of choice is to transfect virus-infected cells with targeting vectors carrying a vaccinia promoter/foreign gene expression unit within a vaccinia homology region, allowing the insertion of foreign DNA by homologous recombination.

Since poxviruses encode their own transcriptional apparatus to allow cytoplasmic transcription, endogenous poxvirus promoters must be used to drive the expression foreign genes, at least in simple vectors. A number of vaccinia promoters have been used, and the gene expression parameters depend upon whether early, late or constant promoters are chosen. Early expression may be desirable to avoid the cytopathic effects of the virus, but the highest levels of transcription are provided by late promoters. The P11 promoter is extensively used, and can generate over 1 mg of protein per 10^6 cells. The constant promoter P7.5 is not quite as active as P11, but allows transcription throughout the replication cycle and is the most widely used of all. The vaccinia virus early transcriptional apparatus uses a specific transcriptional termination signal with the consensus sequence TTTTTNT, so it is advisable to eliminate such motifs within foreign genes to prevent possible truncation. It is also notable that, due to cytoplasmic transcription, vaccinia virus contains no introns and cannot splice introns present in foreign genes. Therefore, vaccinia vectors must be used to express cDNA sequences, or minigenes with introns artificially removed. Higher levels of foreign gene expression can be achieved using a hybrid expression system in which the transgene is driven by the bacteriophage T7 promoter, and the T7 RNA polymerase is expressed from a second vector. High-level transient expression can be achieved if the T7-expressing vector is a plasmid, transiently present in the cell following transfection. More prolonged expression is achieved by incorporating the T7 gene under the control of a vaccinia promoter in a second recombinant virus vector. For toxic proteins, inducible expression systems have also been designed. One example incorporates a strong vaccinia promoter, such as 4b, combined with the *lac* operator sequence from *E. coli*. The foreign gene is co-expressed with *E. coli lacI*, encoding the Lac repressor, allowing foreign gene expression to be regulated by IPTG (inducible expression systems are discussed in detail later).

Vaccinia forms large plaques on permissive cells and these can be lifted onto nitrocellulose or nylon filters and subjected to hybridization-based screening for the foreign gene. The efficiency of screening is enhanced by various selection strategies, which depend on the site of insertion. Negative TK selection is used where the foreign gene is inserted into the viral *Tk* locus: tk- viruses are resistant to the normally lethal effects of 5-bromodeoxyuridine and can be selected on this basis, although naturally occurring tk- mutants are coselected and true recombinants must still be identified by hybridization. Negative HA selection is used where the foreign gene is inserted into the viral hemagglutinin locus: when chicken erythrocytes are added to the plate, HA- plaques are clear, whereas wild-type plaques are red. Selection can also be accomplished by cotransfer of a dominant selectable marker such as neo or gpt, or a visible marker such as *lacZ*. The latter is a popular screening method: recombinant plaques become blue when incubated in the presence of X-gal, while parental plaques remain clear. (See references (43–46).)

4.9   Hybrid viral vectors

We conclude this chapter with a short discussion about the potential of combining certain features of different viruses to make them more suitable vectors. As discussed above, individual viruses have certain advantages and disadvantages as vectors, and none is suitable for all applications. For instance, herpes virus and adenovirus each have a broad host range, and herpes virus in particular has a large capacity, but neither integrates efficiently into the host genome. Conversely, while AAV integrates with great efficiency, it has a limited capacity. Recently, there has been an effort to design customized hybrid viral vectors with selected advantageous properties from each of the component viruses. Although in its infancy, this is one direction in which the field of vector development could expand in the next few years, providing novel and superior vectors for specific applications, especially in the field of gene therapy. Using the example above, a herpes simplex virus vector carrying inducible AAV rep functions could be used to carry recombinant AAV into a broad range of cells, facilitating transfer of the AAV passenger to the genome. The advantage of this strategy is that the AAV genome does not need to be packaged, and there would therefore be no size constraints on the foreign DNA it could carry. (See references (47,48).)

References

1. Berkner KL (1992) Expression of heterologous sequences in adenoviral vectors. *Curr Top Microbiol Immunol* **158**: 39–67.
2. Gerard RD, Meidell RS (1995) Adenovirus vectors. In: Glover DM, Hames BD (eds) *DNA Cloning 4, Mammalian Systems: A Practical Approach*, second edition, pp. 285–306. IRL Press, Oxford, UK.
3. Ring CJA (1996) Adenovirus vectors. In: Lemoine NR, Cooper DN (eds) *Gene Therapy*, pp. 61–76. BIOS Scientific Publishers, Oxford, UK.
4. Douglas JT (2004) Adenovirus mediated gene delivery. An overview. In: Heiser WC (ed) (2004) *Gene Delivery to Mammalian Cells. Volume 2 – Viral Gene Transfer Techniques*, pp. 3–14. Humana Press, Towata, NJ.
5. Curiel DT, Douglas JY (2002) *Adenoviral Vectors for Gene Therapy*. Academic Press, New York.
6. Morsey MA, Caskey CT (1999) Expanded capacity adenoviral vectors – the helper dependent vectors. *Mol Med Today* **5**: 18–24.
7. Imperiale MJ, Kochanek S (2004) Adenovirus vectors: Biology, design, and production. *Curr Top Microbiol* **273**: 335–357.
8. Liu Q, Muruve DA (2003) Molecular basis of the inflammatory response to adenovirus vectors. *Gene Ther* **10**: 935–940.
9. Muzyczka N (1992) Use of adenoassociated virus as a general transduction vector for mammalian-cells. *Curr Top Microbiol Immunol* **158**: 97–129.
10. Bartlett JS, Samulski RJ (1996) Adeno-associated virus vectors for human gene therapy. In: Lemoine NR, Cooper DN (eds) *Gene Therapy*, pp. 77–92. BIOS Scientific Publishers, Oxford, UK.

11. Daly TM (2004) Overview of adeno-associated viral vectors. In: Heiser WC (ed) (2004) *Gene Delivery to Mammalian Cells. Volume 2 – Viral Gene Transfer Techniques*, pp 157–165. Humana Press, Towata, NJ.

12. Flotte TR (2004) Gene therapy progress and prospects: Recombinant adeno-associated virus (rAAV) vectors. *Gene Ther* **11**: 805–810.

13. Lehtonen E, Tenenbaum L (2003) Adeno-associated viral vectors. *Int Rev Neurobiol* **55**: 65–98.

14. Snyder RO, Flotte TR (2002) Production of clinical-grade recombinant adeno-associated virus vectors. *Curr Opin Biotech* **13**: 418–423.

15. Lundstrom K (1997) Alphaviruses as expression vectors. *Curr Opin Biotechnol* **8**: 578–582.

16. Schlesinger S, Dubensky TW Jr (1999) Alphaviruses for gene expression and vaccines. *Curr Opin Biotechnol* **10**: 434–439.

17. Karlsson GB, Liljeström P (2004) Delivery and expression of heterologous genes in mammalian cells using self-replicating alphavirus vectors. In: Heiser WC (ed) (2004) *Gene Delivery to Mammalian Cells. Volume 2 – Viral Gene Transfer Techniques*, pp. 543–557. Humana Press, Towata, NJ.

18. Rayner JO, Dryga SA, Kamrud KI (2002) Alphavirus vectors and vaccination. *Rev Med Virol* **12**: 279–296.

19. Schlesinger S (2001) Alphavirus vectors: development and potential therapeutic applications. *Expert Opin Biol Th* **1**: 177–191.

20. Lundstrom K (2003) Semliki Forest virus vectors for gene therapy. *Expert Opin Biol Th* **3**: 771–777.

21. Patel G, Jones NC (1995) The baculovirus expression system. In: Glover DM, Hames BD (eds) *DNA Cloning 2, Expression Systems: A Practical Approach*, second edition, pp. 205–244. IRL Press, Oxford, UK.

22. O'Reilley DR, Miller LK, Luckow VA (1992) *Baculovirus Expression Vectors: A Laboratory Manual*. Freeman, San Fransisco, CA.

23. King LA, Possee RD (1992) *The Baculovirus Expression System: A Laboratory Guide*. Chapman & Hall, London.

24. Fraser MJ (1992) The baculovirus-infected insect cell as a eukaryotic gene expression system. *Curr Top Microbiol Immunol* **158**: 131–172.

25. Possee RD (1997) Baculoviruses as expression vectors. *Curr Opin Biotechnol* **8**: 569–572.

26. Kost TA, Condreay JP (2002) Recombinant baculoviruses as mammalian cell gene-delivery vectors. *Trends Biotechnol* **20**: 173–180.

27. Merrihew RV, Kost TA, Condreay JP (2004) Baculovirus mediated gene delivery into mammalian cells. In: Heiser WC (ed) (2004) *Gene Delivery to Mammalian Cells. Volume 2 – Viral Gene Transfer Techniques*, pp. 355–365. Humana Press, Towata, NJ.

28. Vos J-MH (1995) Herpesviruses as genetic vectors. In: Vos J-MH (ed.) *Viruses in Human Gene Therapy*, pp. 109–140. Carolina Academic Press/Chapman & Hall, Durham, NC/London, UK.

29. Vos J-MH, Westphal E-M, Banerjee S (1996) Infectious herpes vectors for gene therapy. In: Lemoine NR, Cooper DN (eds) *Gene Therapy*, pp. 127–153. BIOS Scientific Publishers, Oxford, UK.

30. Lim F, Starr P, Song S, Hartley D, Lang P, Wang Y, Geller AI (1995) Expression using a defective herpes simplex virus (HSV-1) vector system. In: Glover DM, Hames BD (eds) *DNA Cloning 4, Mammalian Systems: A Practical Approach*, second edition, pp. 263–283. IRL Press, Oxford, UK.

31. Goins WF, Wolfe D, Krisky DM, Bai Q, Burton EA, Fink DJ, Glorioso JC (2004) Delivery using herpes simplex virus. An overview. In: Heiser WC (ed) (2004) *Gene Delivery to Mammalian Cells. Volume 2 – Viral Gene Transfer Techniques*, pp. 257–298. Humana Press, Towata, NJ.

32. Burton EA, Fink DJ, Glorioso JC (2002) Gene delivery using herpes simplex virus vectors. *DNA Cell Biol* **21**: 915–936.
33. Calderwood NA, White RE, Whitehouse A (2004) Development of herpesvirus-based episomally maintained gene delivery vectors. *Expert Opin Biol Th* **4**: 493–505.
34. Brown AC, Dougherty JP (1995) Retroviral vectors. In: Glover DM, Hames BD (eds) *DNA Cloning 4, Mammalian Systems: A Practical Approach*, second edition, pp. 113–142. IRL Press, Oxford, UK.
35. Miller AD (1992) Retroviral Vectors. *Curr Top Microbiol Immunol* **185**: 1–24.
36. Cepko C (1996) Overview of the retrovirus transduction system. In: Ausubel FM, Brent R, Kingston RE, Moore DD, Seidman JG, Smith JA, Struhl K (eds) *Current Protocols in Molecular Biology*, supplement 17, pp. 9.10.1–9.10.7. John Wiley & Sons, New York.
37. Gunzburg WH, Salmons B (1996) Retroviral vectors. In: Lemoine NR, Cooper DN (eds) *Gene Therapy*, pp. 33–60. BIOS Scientific Publishers, Oxford, UK.
38. Somia N (2004) Gene transfer by retroviral vectors. An overview. In: Heiser WC (ed) *Gene Delivery to Mammalian Cells. Volume 2 – Viral Gene Transfer Techniques*, pp. 463–490. Humana Press, Towata, NJ.
39. Federico M (1999) Lentiviruses as gene delivery vectors. *Curr Opin Biotechnol* **10**: 448–453.
40. Kafri T (2004) Gene delivery by lentivirus vectors. An overview. In: Heiser WC (ed) (2004) *Gene Delivery to Mammalian Cells. Volume 2 – Viral Gene Transfer Techniques*, pp. 367–390. Humana Press, Towata, NJ.
41. Blesch A (2004) Lentiviral and MLV based retroviral vectors for ex vivo and in vivo gene transfer. *Methods* **33**: 164–172.
42. Sandrin V, Russell SJ, Cosset FL (2003) Targeting retroviral and lentiviral vectors. *Curr Top Microbiol* **281**: 137–178.
43. Moss B (1992) Poxvirus expression vectors. *Curr Top Microbiol Immunol* **185**: 25–39.
44. Carroll MW, Moss B (1997) Poxviruses as expression vectors. *Curr Opin Biotechnol* **8**: 573–577.
45. Mackett M (1995) Construction and characterization of vaccinia virus recombinants. In: Glover DM, Hames BD (eds) *DNA Cloning 4, Mammalian Systems: A Practical Approach*, second edition, pp. 43–83. IRL Press, Oxford, UK.
46. Guo ZS, Bartlett DL (2004) Vaccinia as a vector for gene delivery. *Expert Opin Biol Th* **4**: 901–917.
47. Reynolds PN, Feng M, Curiel DT (1999) Chimeric viral vectors – the best of both worlds. *Mol Med Today* **5**: 25–31.
48. Lam PYP, Breakefield XO (2000) Hybrid vector designs to control the delivery, fate and expression of transgenes. *J Gene Med* **2**: 395–408.

7. The next morning, prepare a 1.3% mix of Noble agar in water, autoclave to sterilize and allow to set. Re-melt the agar in a microwave and allow it to cool to 45°C in a water bath. Meanwhile prepare a solution of 2× MEM medium + 4% fetal bovine serum, 100 µg ml^{-1} penicillin, 20 µg ml^{-1} streptomycin and 50 µg ml^{-1} Fungizone. Prewarm to 37°C. Mix the agar and MEM medium 1:1 and swirl. Allow to stand while removing the medium from the cells, and then gently add 6 ml of the agar/MEM mix to each dish. Swirl to mix the agar/MEM with residual medium. Allow gel to solidify at room temperature (5–10 minutes) and then return cells to incubator.

8. Incubate for up to 3 weeks, checking every few days for the appearance of viral plaques representing clear areas under the agar. Each week, prepare a fresh mix of agar/MEM and add 3 ml to each dish.

9. Harvest recombinant virus by picking plugs of agar above well-separated plaques using a Pasteur pipette and deposit the plugs into Eppendorf tubes containing 1 ml of DMEM + 2% fetal bovine serum. Freeze thaw each tube once.

10. For the small-scale propagation of recombinant virus stocks and the production of small amounts of recombinant protein for testing, seed 100 mm dishes with 293 cells and allow them to reach 90% confluence. Then add the viral lysate prepared in step 9 and incubate the cells at 37°C.

11. When most of the cells show signs of infection, freeze–thaw the cells to lyse them and harvest the medium.

12. Centrifuge at 13 000 g for 10 min at 4°C to remove cell debris and retain the supernatant, which should contain the viral stock and any recombinant protein produced in the infection.

13. The viral stock can be used to scale up production if desired by the infection of fresh dishes of 293 cells. Large amounts of virus can be collected from the lysed cells and purified by chromatography or ultracentrifugation.

14. The purified viral stocks can then be used to infect and transduce other cell lines and facilitate transient transgene expression, although productive viruses will not be produced from such cells because they lack the helper functions provided by 293 cells.

Protocol 4.2: Transduction of cultured cells with MLV-based retroviral vectors

1. Helper-free stocks of recombinant MLV are produced by transfecting packaging cells with a DNA copy of the vector contained in a plasmid (since the MLV genome is small, many plasmids are available in which unique restriction sites are available for the insertion of cloned genes). On the day before transfection, plate out the packaging cells at a density of 1.5×10^5 per well in a six-well plate and incubate overnight.

2. Transfection can be achieved using any of the procedures described in Chapter 2, although the lipofection procedure is preferred. Prepare 5 μg of vector DNA for each lipofection. Co-transfection with a plasmid containing a marker gene (e.g. *neo*) allows the selection of stably transformed cells, which produce higher viral titers than transiently transfected cells, but do not use a marker which has been used to construct the packaging line!

3. When transfection is complete, rinse the cells with DMEM and then refresh with DMEM + 10% fetal calf serum + the selection reagent (e.g. G418 400 μg ml^{-1}) at the appropriate concentration and refresh daily, splitting the cells 1:10 into fresh selective medium when they become confluent.

4. When ready to harvest the virus, allow the virus producing cells to reach near confluence.

5. Replace the medium with half the normal volume, lacking selection agents and incubate for a further 24 hours before harvesting the medium. Alternatively, harvest the medium after 12 hours, refresh the cells and repeat the procedure after another 12 hours.

6. Transfer the harvested medium to a suitable container and centrifuge at 5000 *g* for 5 min to remove dead cells and other debris. The medium can also be filtered to remove debris (use a 0.45 μm filter).

7. To determine the titer of the viral stock, plate NIH 3T3 cells at 1.5×10^5 per well in a six-well plate one day before infection and then make serial dilutions of the viral stock in DMEM supplemented with 8 μg ml^{-1} polybrene (this reagent facilitates the interaction between viral particles and the plasma membrane of target cells).

8. Remove the medium from the cells and replace with 0.5 ml of virus stock in DMEM. Incubate the cells at 37°C for 2 h.

9. Remove the medium and replace with 3 ml of fresh medium. Return to the incubator for 48 h.

10. For vectors carrying a selectable marker, split the cells at a ratio of 1:10 and 1:30 into fresh medium containing the selection agent at the appropriate concentration (e.g. for *neo* use 400 mg ml^{-1} G418). Incubate for 10 days, changing the medium every 3 days. At the end of the incubation period, count the number of resistant colonies.

11. For vectors carrying a visible marker such as *lacZ*, fix the cells in 2% paraformaldehyde in PBS of 15 min at room temperature, rinse several times with PBS and then incubate the cells with the appropriate substrate for reporter molecule detection. For *lacZ*, use X-gal staining solution and count the number of blue-stained cell clones.

12. Use the results from steps 10 or 11 to determine the titer of the vector, remembering to consider the dilution of the original stock and (in step 10) the ratio in which the cells were split. For efficient transduction, titers of 10^5–10^7 are optimal and the supernatant (or filtrate) from step 6 can be used directly to infect target cells. If the titer is lower, the stock can be concentrated up to 50-fold by centrifugation at 30 000 *g* for 16 h at 4°C and resuspension in the residual volume once most of the supernatant has been removed.

13. For transduction of other target cells, repeat step 7 using the appropriate target cells in place of NIH 3T3 cells.

Genetic manipulation of animals

5

5.1   Introduction

In the previous three chapters we have considered the available methods for transferring genes into animal cells. In this chapter we advance the gene transfer principle a little further and discuss the genetic modification not of single cells, but of *whole animals*. The ability to manipulate animal genomes is one of the most significant scientific advances in the last 50 years, providing the ultimate assay system for the investigation of gene function and regulation. Furthermore, and as discussed in more detail in Chapter 10, genetically modified animals have many practical applications: they can provide models of human diseases, they can be used as bioreactors to produce valuable proteins, they can be used to test novel drugs and therapies, and it is hoped ultimately that animals will be used to grow replacement organs that can be used in human beings. The genetic modification of domestic species can also be used to improve quality traits in farm animals, increase resistance to disease and generate elite herds. An animal in which every cell carries new genetic information is described as *transgenic*. Originally, this term was restricted to animals carrying foreign DNA (i.e. DNA from a different species, or recombinant DNA) but nowadays it is quite acceptable to describe an animal carrying additional copies of endogenous sequences as transgenic as long as those sequences have been introduced into the genome artificially as exogenous DNA. However, most scientists do not describe genetically modified animals as transgenic if they carry very subtle mutations, such as point mutations introduced by gene targeting (see Chapter 6).

We must distinguish right at the beginning of this chapter the difference between a transgenic animal and an animal (or human) that has undergone *in vivo* DNA transfer using the methods described in Chapters 2–4. The hurdle that must be overcome when producing transgenic animals is the introduction of the same exogenous DNA sequence at the same locus in every cell *including the germline cells* so that the transgenic animal produces transgenic offspring. Simply introducing DNA into the bloodstream or into a specific population of skin or muscle cells will not achieve this, which is why gene therapy targeted at somatic tissues in humans is considered ethically acceptable while germline gene therapy is not. Unlike the situation in a dish of cultured cells, it is not possible simply to introduce DNA into an animal and select those cells that have been transformed. Instead, DNA must be introduced into single cells that have the potential to produce an entire animal, such cells being described as *totipotent* or *pluripotent*[1]. This generally involves introducing DNA directly into the germ cells, gametes or early embryo (prior to the development of the germline), although in some strains of mice it is possible to exploit the advantages of cell culture through the use of embryonic stem (ES) cells, which are derived from the preimplantation embryo and can contribute to all the tissues of the developing animal (including the germline) if introduced into a host embryo at the correct developmental stage. ES cells also undergo homologous recombination with relatively high efficiency, and we discuss this property further in the next chapter, which considers the principles of gene targeting. The methods for generating germline-transformed animals are summarized in *Table 5.1*.

Although animal cells become progressively restricted during development, the nuclei of at least some somatic cells remain totipotent even in

Table 5.1 Gene transfer to animals: Summary of targets and methods for transforming the germline

Target cells	Method
Germ cells	Transfection of cultured primordial germ cells (mammals) Injection into embryo at site of germ cell development (*Drosophila*)
Sperm	Attachment of DNA to sperm heads (mammals) Introduction of DNA into decondensed sperm nuclei (*Xenopus*)
Egg/zygote	Microinjection into egg cytoplasm (birds, amphibians, fish, *C. elegans*) Pronuclear microinjection (mammals) Retroviral transfer (birds and mammals incl primates) Transfer of genetically modified somatic or ES cell nucleus to enucleated egg Electrofusion with enucleated somatic cell containing genetically modified mitochondria (mitochondrial transgenics)
Blastocyst	Microinjection into blastocele (mammals) ES cell transfer (mice) Retroviral transfer (mammals, birds)
ES cells (followed by cell transfer to blastocyst, or nuclear transfer to egg)	Transfection – transgene addition Transfection – gene targeting Retroviral transduction
Somatic cells (followed by nuclear transfer to egg)	Transfection – transgene addition Transfection – gene targeting Retroviral transduction

1 These terms are not synonymous. *Totipotent* cells such as eggs have the capability to produce *all* the cell types of the developing organism, and in the case of mammals this would include all the extra-embryonic membranes. *Pluripotent* cells are more restricted, but the most potent of them can still produce all the cell types in the embryo. Murine ES cells are pluripotent – they can contribute to all cells in the embryo, including the germline, but they have lost the capacity to produce cells of the extra-embryonic membranes.

the adult. For many years it has been known that amphibians can be cloned by transferring nuclei from adult cells into enucleated eggs, and in 1997 the same principle was demonstrated in mammals, resulting in the now famous sheep named Dolly. Since then, at least ten further species of mammals have been cloned by nuclear transfer from adult cells, and the first albeit unconfirmed reports of human cloning have been published. Since nuclear transfer represents a rapid way to scale up herds of transgenic or genetically modified animals, we also discuss the principles and methods for nuclear transfer in this chapter.

5.2   Historical perspective – of mice and frogs … and flies

The first transgenic animals were mice, sporadically reported throughout the 1970s. For example, Jaenisch and Mintz (1) showed that mice could be transformed by injecting SV40 virus DNA into the blastocele cavities of preimplantation embryos, which were then implanted in foster mothers and allowed to develop to term. Some of the resulting embryos carried integrated SV40 DNA, and in a very few cases this was transmitted to the subsequent generation showing that germline transformation had been achieved. However, the first reliable method for the production of transgenic mice was reported in 1980 by Brinster and colleagues (2), and involved a technique known as pronuclear microinjection in which DNA is introduced into the male pronucleus of the recently fertilized egg (see below). This followed a series of experiments involving the injection of mRNA into eggs and zygotes at various stages of development, which had demonstrated that exogenous RNA could be translated. Similar experiments had been carried out using another very popular experimental model organism, the African clawed frog, *Xenopus laevis*. In this species, both DNA and mRNA were expressed so efficiently that *Xenopus* oocytes soon came to be a widely used heterologous expression system (3). However, unlike the situation in the mouse, DNA integration was extremely rare. Although transgenic *Xenopus* were documented as early as 1981 (5), a reliable system for transgenesis in the frog was not developed until 1996 (6). This delay partly reflects the long generation time of *Xenopus laevis*, which meant that few researchers were interested in developing transgenic lines, but primarily results from the fact that amphibian and mammalian eggs are very different in developmental terms and react in different ways to the presence of exogenous DNA. This is discussed in greater detail in Section 5.7.

At about the same time, Spradling and Rubin had developed an efficient system for the transformation of another popular model organism, the fruit fly *Drosophila melanogaster*. As in frogs, DNA introduced into insect eggs integrates into the genome very inefficiently. However, by exploiting a transposon called the P-element, Spradling and Rubin produced the first transgenic flies in 1982 (7). The P-element has the useful ability of jumping from site to site in the genome, and can likewise jump from a plasmid into the genome if the plasmid is introduced by microinjection into the germ cell region of the *Drosophila* egg. By manipulating the P-element in the laboratory and introducing a cargo of exogenous DNA, flies carrying any transgene of interest could be produced (Section 5.9). (See references (1–7).)

5.3 Standard methods for the production of transgenic mice

Pronuclear microinjection

The technique that has become established for the production of transgenic mice is the injection of DNA into the male pronucleus, as shown in *Figure 5.1*. Immediately after fertilization, the sperm nucleus is uncoated but remains separated from the smaller female pronucleus until after the first division of the zygote. At this stage, a small amount of DNA can be introduced through a fine needle, prepared by stretching a glass pipette in a Bunsen flame. During the transfer procedure, the egg is held in place with a suction pipette. After injection, the embryos are cultured in vitro until the 8- or 16-cell stage, and then transferred to females that have been

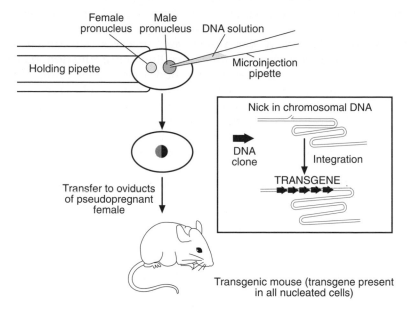

Figure 5.1

Construction of transgenic mice by pronuclear microinjection. Very fine glass pipettes are constructed using specialized equipment: one, a holding pipette, has a bore which can accommodate part of a fertilized oocyte, and thereby hold it in place, while the microinjection pipette has a very fine point which is used to pierce the oocyte and then the male pronucleus (because it is bigger). An aqueous solution of the desired DNA is then pipetted directly into the pronucleus. The introduced DNA clones can integrate into chromosomal DNA at nicks, forming transgenes, usually containing multiple head-to-tail copies. Following withdrawal of the micropipette, surviving oocytes are reimplanted into the oviducts of pseudopregnant foster females (which have been mated with a vasectomized male; the mating act can initiate physiological changes in the female which stimulate the development of the implanted embryos). Newborn mice resulting from development of the implanted embryos are checked by PCR for the presence of the desired DNA sequence. Taken from Strachan *et al*, *Human Molecular Genetics*, Copyright (2004), Garland Science.

rendered pseudopregnant by mating with vasectomized males, for development to term (*Protocol 5.1*).

The DNA introduced by microinjection tends to integrate at a single locus (although multiple integration sites have been documented) generally as an array of 10–50 copies. Higher copy numbers have been reported, and this may be a function of the amount of injected DNA. In many cases, the DNA integrates immediately and the resulting F_0 generation mouse (primary transformant or founder) is transgenic. In other cases, the DNA may not integrate until one or two rounds of division have been completed, so the early embryo is mosaic (a mixture of transformed and non-transformed cells derived from the same genetic background) rather than transgenic. Under such circumstances, the founder mouse itself may still be transgenic, if all the non-transformed cells contribute to the extra-embryonic membranes, but is more likely to be mosaic. It is necessary to test the first generation of offspring (F_1) for the presence of the transgene to confirm germline transmission. It is possible that the germline itself is mosaic, which means that only a proportion of the offspring will carry the transgene.

The pronuclear microinjection technique is reliable, varying in efficiency from 5% to 50% (proportion of transgenic lines from injected eggs). Occasionally, there may be large-scale genomic rearrangements or deletions at the integration site, and in some cases the entire locus may be hyper-methylated and silenced (Chapter 8). The site of integration is random which probably reflects the position of pre-existing chromosome breaks.

Transduction with recombinant retroviruses

As discussed in Chapter 4, recombinant retroviruses provide a natural mechanism for stably introducing DNA into the genome of dividing animal cells. Retroviruses can infect early mouse embryos as well as cultured ES cells (see below), so recombinant retroviral vectors have been widely used for mouse transgenesis (*Figures 5.2* and *5.3*). While it is true that retroviral transduction tends to generate single copy transgenes without significant disruption of the surrounding genomic DNA, there are several major limitations which restrict the usefulness of this method. These include the limited capacity of retroviral vectors such that large transgenes cannot be integrated in this manner, the tendency of viral regulatory elements to interfere with the expression of surrounding endogenous genes, and the tendency for the integrated provirus to become methylated and silenced. Founder embryos are always mosaics rather than transgenic, so germline transmission has to be confirmed in the F_1 generation. Nowadays, retroviral transduction is rarely used to generate transgenic mice because of the greater benefits of other methods. However, it remains useful for gene transfer to birds and other mammals (see below) and as a method for generating transgenic sectors of embryos for developmental studies, e.g. in limb development.

Transfection of ES cells

ES cells are derived from the inner cell mass of the preimplantation mouse embryo. They have a combination of four advantages which makes them

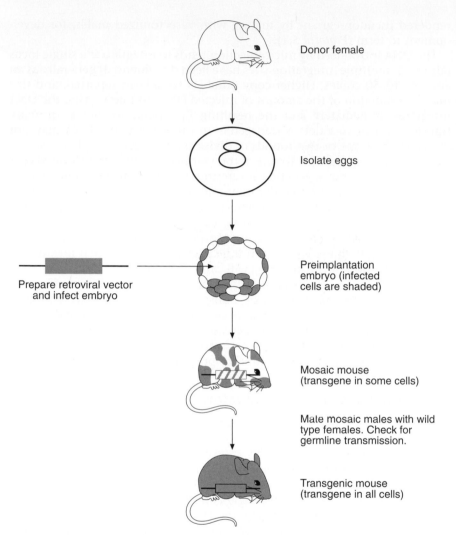

Donor female

Isolate eggs

Prepare retroviral vector
and infect embryo

Preimplantation
embryo (infected
cells are shaded)

Mosaic mouse
(transgene in some cells)

Mate mosaic males with wild
type females. Check for
germline transmission.

Transgenic mouse
(transgene in all cells)

Figure 5.2

Construction of transgenic mice by retroviral transduction. Retroviral vectors can
be used to infect the preimplantation embryo, leading to the possible integration
of single transgene copies into germline precursors. The resulting mice are mosaic
for retroviral integration, but may give rise to transgenic mice in the following
generation.

unique and of particular practical value for the genetic manipulation of
mice. First, they are pluripotent, i.e. they can contribute to all cells of the
developing embryo proper, including the germline (they are not totipotent
because they cannot form cells of the extra-embryonic membranes other
than the amnion). Second, they can be cultured in vitro in the same
manner as any immortalized cell line, and they can be transfected or transduced using all the standard methods described in Chapters 2 and 4. Third,

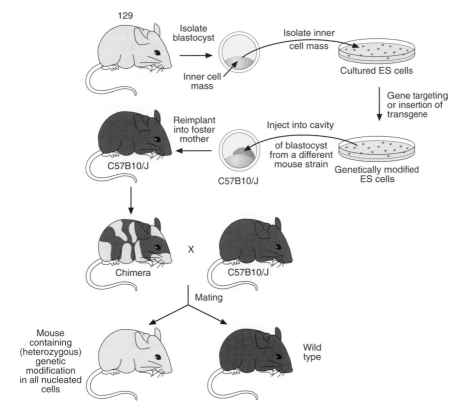

Figure 5.3

Genetically modified ES cells for transferring foreign DNA or specific mutations into the mouse germline. Cells from the inner cell mass are cultured following excision of oviducts and isolation of blastocysts from mouse strain 129. These ES cells retain the capacity to differentiate into all the different types of tissue in the adult mouse. ES cells can be genetically modified while in culture by insertion of foreign DNA or by introducing a subtle mutation (Chapter 6). The modified ES cells can then be injected into isolated blastocysts of another mouse strain (e.g. C57B10/J which has a black coat color that is recessive to the agouti color of the 129 strain) and then implanted into a pseudopregnant foster mother of the same strain as the blastocyst. Subsequent development of the introduced blastocyst results in a chimera containing two populations of cells (including germline cells), which ultimately derive from different zygotes (normally evident by the presence of differently colored coat patches). Backcrossing of chimeras can produce mice that are heterozygous for the genetic modification. Subsequent interbreeding of heterozygous generates homozygotes. Taken from Strachan *et al*, *Human Molecular Genetics*, Copyright (2004), Garland Science.

and as will be discussed in more detail in Chapter 6, they are unusually amenable to homologous recombination, which means they can be used for gene targeting. Finally, they can be induced to differentiate in a variety of different ways by adding specific growth factors and other components to the culture medium, or they can be maintained indefinitely in an undifferentiated state. All together, this means that ES cells can be transformed using a variety of methods including standard transfection and transduction techniques but also specialized methods such as microcell-mediated transfer for the introduction of large DNA segments. The transformed cells can be selected using standard marker genes, whereas transgene integration in injected eggs and embryos can only be confirmed by the molecular analysis of the resulting mice. The pre-selected transformed ES cells can be introduced into another mouse embryo and can contribute to all tissues, including the germline.

The colonization of a host blastocyst with donor ES cells creates a chimeric embryo[2] and this allows the contribution of ES cells to the germline to be established using visible markers (*Figure 5.3*). Most common ES cell lines are derived from male mice of strain 129, which has the dominant coat color known as agouti. The normal approach is to use host embryos from a mouse strain such as C57BL/6J, which has a recessive black coat color. Colonization of the embryo by ES cells generates predominantly male chimeras with patchwork coats of black and agouti cell clones. If the ES cells have contributed to the germline, mating chimeric males with C57BL/6J females generates heterozygous transgenic offspring with the agouti coat color, confirming germline transmission. (See references (8–13).)

5.4 Further methods for mammalian and avian transgenesis

The three standard methods applied in other animals

The three methods for producing transgenic mice discussed above have also been used in other mammals and birds, but the efficiency is much lower. Pronuclear microinjection in domestic mammals typically results in a transgenesis rate of <1% and this problem is exacerbated by the greater difficulty and expense of obtaining eggs from donor females. The eggs themselves are delicate, making them more difficult to handle, and also tend to be opaque which makes it more difficult to see the target nuclei without first centrifuging the eggs to clear the cytoplasm. It is possible to remove chicken eggs after fertilization and microinject DNA into the cytoplasm of the germinal disc, where the male and female pronuclei are to be found, but manipulated eggs have to be cultured *in vitro* because it is impossible to re-implant them. Love *et al.* (16) were able to produce seven chicks that survived to adulthood using this method, and one of the cockerels transmitted the transgene to a small proportion of his offspring, indicating he was a mosaic for transgene integration.

Retroviruses have also been used to produce transgenic chickens following the injection of a replication-defective reticuloendotheliosis vector into

---

2 Founder mice are often mixtures of transformed and non-transformed cells, which may be described as either mosaics or chimeras. The terms are not synonymous. A *mosaic* founder mouse contains a mix of transformed and non-transformed cells, which are derived from the same genetic background (e.g. embryonic cells derived from the same egg, only some of which have been transduced by a retrovirus). A *chimeric* founder mouse contains a mixture of transformed and non-transformed cells from different genetic backgrounds (e.g. host embryonic cells and donor ES cells).

laid eggs. About 10% of the resulting cockerels carried the transgene and 20 transgenic lines were generated. Most retroviruses can only infect dividing cells because they gain entry to the nucleus when the nuclear envelope breaks down. The nuclear envelope also breaks down during meiosis, allowing the use of retroviruses to transfer exogenous genes into eggs at around the time of fertilization. This has been achieved in cattle following the injection of replication-defective retroviral vectors into the perivitelline space of isolated bovine oocytes during the second meiotic division and, most recently, in rhesus monkeys[3]. Some of the disadvantages of standard retroviral vectors, particularly their requirement for dividing cells and the tendency to undergo silencing, have been addressed more recently by the development of techniques for lentiviral transgenesis, which has been achieved in mice, rats, pigs and cattle.

The standard transgenesis method which has become the most useful in mice, i.e. the transfection or transduction of ES cells, has been the most difficult to reproduce in other animals. It has proven consistently impossible to derive true ES cells from any mammalian species other than mice, with the exception of humans, although chicken ES cells have been available since 1999 (25). Instead, attention has turned towards the possible transfection of cultured primordial germ cells (PGCs), which are the embryonic cells that eventually produce gametes. PGCs can be transfected directly or they can be cultured and derivatized to produce embryonic germ cells (EGCs) that appear very similar in morphology to ES cells. Chicken PGCs have been isolated, transduced with retroviral vectors and replaced in the germline, yielding mosaic birds that showed germline transmission of the transgene; this has now become one of the most straightforward means of gene transfer to poultry. Although less established, PGC transfer has also been achieved in mice, rabbits, pigs and cattle. Transgenic mice and pigs have also been produced by an alternative technique in which immature sperm cells (spermatogonia) are transfected in vitro and then implanted into the testis to produce transgenic sperm. (See references (14–28).)

Other methods – intracytoplasmic sperm injection

Intracytoplasmic sperm injection (ICSI) is the injection of sperm heads directly into the cytoplasm of the egg, a technique which is offered to some infertile couples. The technique has been adapted for gene transfer following the demonstration that murine sperm heads can bind spontaneously to naked plasmid DNA in vitro, resulting in the highly efficient introduction of a green fluorescent protein transgene into mouse eggs (*Figure 5.4*). Indeed, following the injection of plasmid-coated sperm into unfertilized oocytes, nearly 95% of the developing mouse embryos showed transient GFP activity and about 20% were shown to be transgenic. The same method has more recently been used to generate transgenic pigs but it was not possible to produce transgenic Rhesus monkeys, despite strong GFP activity into the blastula stage of development. Variations on this method have

3 The first transgenic monkey was generated by retroviral gene transfer and was named ANDi, apparently because when the letters are reversed they spell iDNA (inserted DNA).

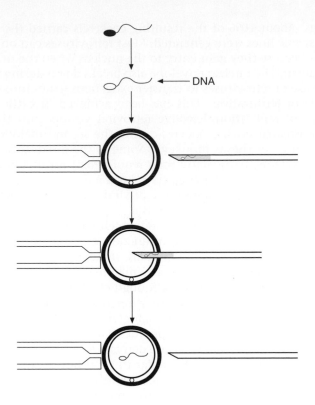

Figure 5.4

Principles of gene transfer by intracytoplasmic sperm injection. Sperm are isolated and mixed with a DNA solution, resulting in the spontaneous binding of plasmid DNA to the sperm head. Individual sperm are then captured in a micropipette and injected into recipient eggs, held in place with a suction pipette. The plasmid DNA diffuses from the sperm head as the sperm is deconstructed and the nucleus decondensed, resulting in transgene integration in about 20% of the nuclei.

used lipoplexes, antibodies and electroporation in attempts to increase the efficiency of sperm–DNA binding. (See references (29–31).)

5.5 Mitochondrial transgenics and electrofusion

The techniques discussed above have all been developed for nuclear transgenesis, i.e. the integration of DNA into the nuclear genome. However, a number of human genetic diseases are caused by mitochondrial mutations, and to model such diseases it is necessary to introduce DNA into the mitochondrial genome of suitable animals. This cannot be achieved using standard transgenesis techniques and a novel method has been developed in which donor cell lines carrying mitochondrial mutations are enucleated and then fused with recipient mouse zygotes. The resulting *transmitochondrial mice* carry the mutation in every cell and pass the mutation through the female germline, since mitochondria in the sperm only rarely contribute to the embryo.

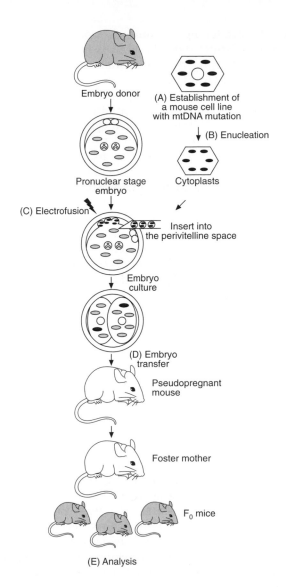

Figure 5.5

Electrofusion procedure for producing mitochondrial transgenic mice with exogenous mitochondrial DNA derived from a donor cell line. The steps in the procedure include: (A) establishing a cell line carrying a mitochondrial mutation, often by fusion of mitochondria-depleted cells with synaptosomes or cells containing known mitochondrial mutations; (B) enucleation of that 'cybrid' cell line to produce cytoplasts; (C) insertion of cytoplasts under the zona pellucida of a recipient mouse egg followed by electrofusion to generate a cybrid embryo; (D) culture of the embryo and transfer to a pseudopregnant female; and (E) molecular analysis of the resulting pups. Reprinted from Methods, Vol 26, No.4 , Inoue *et al*, "*Production of Mitochondrial . . .*", Pages 358–363, Copyright (2002), with permission from Elsevier.

The procedure for producing transmitochondrial mice is shown in *Figure 5.5* and involves five distinct steps. First, a cell line must be produced carrying mitochondria containing the appropriate transgene or mitochondrial mutation. There are many ways to achieve this, but one of the most efficient is fusion between cells depleted in mitochondrial DNA (mtDNA) and synaptosomes from aged mice, which often contain mtDNA deletions and other mutations. The resulting cell lines are called cybrids because their cytoplasm is a hybrid of two cell lines. The next step is to remove the nucleus from the cybrid cells, so that they can act as mitochondrial donors without introducing additional nuclei into the egg. This is generally achieved by centrifugation in the presence of cytochalasin B. After washing the enucleated cybrids, zygotes are collected from donor females and a micromanipulator is used to drill a hole through the zona pellucida and place the cybrids in the perivitelline space. After a brief recovery period in culture, embryos containing cybrids are washed in a medium that promotes fusion and placed between the electrodes of a fusion chamber. A long AC pulse (225 V, 10–15 s) is used to align the cytoplasts and oocytes, and then a brief DC pulse (2500 V, 20 ms) is used to induce fusion, which occurs within the next hour. After a brief recovery period in culture, the fused embryos are implanted into pseudopregnant foster mothers and raised to term. (See references (32,33).)

5.6 Nuclear transfer technology

Over 50 years ago, Briggs and King (34) established the principle of *nuclear transfer* in amphibians by removing nuclei from the cells of a blastula stage embryo of the frog *Rana pipens* and injecting them into an enucleated egg. Although few of the resulting cloned embryos survived to adulthood, large numbers of cloned tadpoles could be produced showing in principle that the nucleus of a differentiated cell retained all the information necessary to recapitulate development. Some years later, similar results were obtained using *Xenopus laevis*, following UV irradiation of the recipient egg. Such experiments showed that although animal cells tend to become progressively restricted in their developmental potential as development proceeds, the nuclei often remain pluripotent and can be *reprogrammed* by the cytoplasm of the egg allowing them to start the developmental program all over again.

In all species examined to date, the efficiency of nuclear transfer appears to diminish as development proceeds showing that the nucleus of a differentiated cell does undergo some changes, but these are not irreversible. Thus, in the 1990s it was established that nuclear transfer was also possible in mammals, but for quite some time only nuclei from very early embryos appeared compatible with the procedure. As in the mitochondrial transfer technique discussed above, the transfer of new genetic information into the egg was achieved by electrofusion, although since in this case the aim was nuclear rather than mitochondrial transfer, the donor embryonic cell remained nucleated and the recipient egg was enucleated (usually by simply removing the nucleus with a pipette).

The late 1990s saw a resurgence in interest in mammalian cloning when the first cloned sheep were produced by nuclear transfer from a cultured cell line. Interestingly, while gene transfer to mice has always been technically much simpler than in other animals, quite the reverse applies in nuclear transfer technology where domestic mammals are among the easiest to manipulate. In 1997, Wilmut and colleagues (35) reported the birth of Dolly the sheep (*Figure 5.6*), the first mammal cloned by nuclear transfer from an adult cell (a mammary epithelial cell line). A critical factor in this procedure was the quiescent state of the donor cells, brought about by serum withdrawal, which allowed synchronization between the donor and recipient cell cycles (*Figure 5.7*). Even so, the cloning efficiency was <0.5%; only one of 250 transfers produced a viable lamb.

Since 1997, nuclear transfer from adult cells has resulted in the successful cloning of many further mammalian species. After sheep, came mice, cows,

Figure 5.6

Dolly, the first mammal produced by cloning from an adult cell, together with her lamb Bonnie. Copyright (1998), The Roslin Institute. Used with permissions.

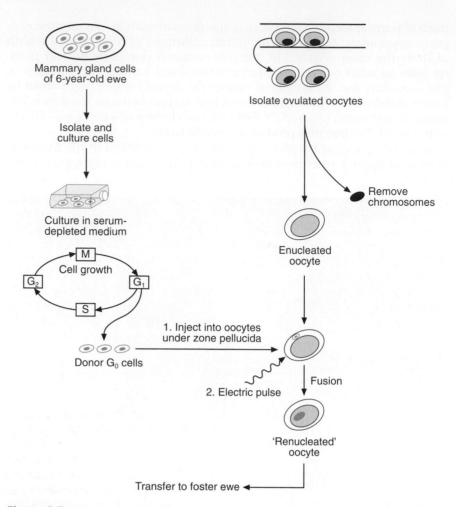

Figure 5.7

The experimental strategy used by Wilmut *et al.* (35). The donor nuclei were derived from a cell line established from adult mammary gland cells. Nuclear transfer was accomplished by fusing individual somatic cells to enucleated, metaphase II-arrested oocytes. The donor cells were deprived of serum before use, forcing them to exit the cell cycle and enter a quiescent state known as G_0 where only minimal transcription occurs. Since eggs are normally fertilized by transcriptionally inactive sperm whose nuclei are presumably 'programmed' by transcription factors and other chromatin proteins available in the egg, the G_0 nucleus may represent the ideal basal state for reprogramming. Note that in other cloning experiments, different strategies have been used to introduce the donor nucleus into the egg. For example, in the successful cloning of adult mice reported by Wakayama *et al.* (37), a very fine needle was used to take up the donor cell nucleus with minimal contamination by donor cell cytoplasm. The donor cell was quickly, but very gently, microinjected into the enucleated oocyte. Taken from Strachan *et al, Human Molecular Genetics*, Copyright (2004), Garland Science.

pigs and goats, with cloning efficiencies ranging from under 1% in pigs and 1–3% in mice to up to 4% in sheep, 5% in cows and over 7% in one study using goats. More recent successes include a cloned cat named CC (for Copy Cat), a cloned rabbit, cloned horses, mule, deer, oxen and rats. Despite major efforts, there has been no success thus far in the production of a cloned primate although various individuals, nations and religious sects have put forward as yet unsubstantiated claims to have produced the first cloned human being. There are also several ongoing projects looking at the feasibility of cloning rare animals or animals in captivity representing species that are extinct in the wild.

The success of nuclear transfer in domestic mammals provides an additional strategy route for the production of transgenics, i.e. gene transfer into the cultured cells that are subsequently used as the source of donor nuclei for nuclear transfer. This approach shares certain advantages with the use of ES cells, including the ability to screen transformed cells in culture for transgene integration and, if necessary, high-level transgene expression prior to the nuclear transfer step. The production of a transgenic mammal by nuclear transfer from a transfected cell line was achieved first by Schnieke *et al.* (44), who introduced the gene for human factor IX into fetal sheep fibroblasts and transferred the nuclei to enucleated eggs. The resulting sheep, Polly, produced the recombinant protein in her milk and can therefore be used as a bioreactor (Chapter 10). McCreath *et al.* (45) later succeeded in producing a transgenic sheep by nuclear transfer from a somatic cell whose genome had been specifically modified by gene targeting (see Chapter 6). More recently, several cloned transgenic cows have been produced, e.g. cows producing human antibodies, as well as cloned transgenic goats and pigs, and cloned pigs generated by gene targeting (Chapter 6). (See references (34,35,44,45).)

5.7 Transgenic frogs

As discussed earlier, DNA introduced into frog embryos tends to remain episomal rather than integrating into the genome, making it difficult to produce transgenic lines. Indeed, while extrachromosomal DNA in mammalian cells is broken down quite rapidly, DNA introduced in frog eggs not only persists well into tadpole development, but even replicates, increasing up to 100-fold by the gastrula stage. In later development, the amount of episomal DNA per embryo decreases, and most of the remaining DNA appears to integrate, although in a highly mosaic fashion representing different integration events in different cells. This difference between mammals and amphibians probably reflects their distinct developmental strategies: slow and asynchronous cleavage divisions in mammals with the early onset of embryonic gene expression, compared with rapid and synchronous divisions in *Xenopus*, the absence of embryonic gene transcription, and hence the necessity for large amounts of maternal mRNA and protein to be stored in the egg. DNA replication in amphibian embryos relies on stored maternal gene products, so there is a stockpile of chromatin assembly proteins and replication enzymes. Exogenous DNA injected into *Xenopus* eggs is therefore assembled immediately into chromatin and undergoes replication in tune with the rapid DNA synthesis already occurring in the nucleus.

Because of the tendency for exogenous DNA to remain in an episomal state in frog eggs, an efficient process for *Xenopus* transgenesis was developed involving the manipulation of the sperm nucleus (6). In this technique, known as restriction enzyme-mediated integration (REMI), linearized plasmids containing the transgene of interest are mixed with decondensed sperm nuclei in a highly artificial environment, and treated with limiting amounts of a restriction enzyme to introduce nicks in the DNA (*Figure 5.8*). The nuclei are then introduced into unfertilized *Xenopus* eggs, where the DNA is repaired, resulting in the integration of plasmid DNA into the genome. The decondensed nuclei are very fragile and require careful handling. The technique generates transgenic embryos which generally survive until the tadpole stage, but is also sufficient for the production of transformed adults which found transgenic lines. As well as *Xenopus laevis*, which is a tetraploid species, the technique has also been applied in *Xenopus tropicalis*, which is diploid and has a shorter generation interval. (See references (6,46).)

5.8 Transgenic fish

At the time of writing, the number of reported species of transgenic fish – nearly 40 – is, perhaps surprisingly, much larger than the number of transgenic mammals. The list includes model species such as the zebrafish (*Danio rerio*), puffer fish (*Fugu rubripes*) and medaka (*Oryzias latipes*), commercially important food stocks (e.g. salmon and trout) and many other species such as goldfish, pike, loach, tilapia and catfish. Since 1985 when the first transgenics (goldfish) were reported, gene transfer in fish has predominantly been achieved by injection of DNA into the cytoplasm of the fertilized egg, generally resulting in mosaic integration and the production of transgenics in the F1 generation. The efficiency of this procedure is comparable with DNA transfer to domestic mammals, i.e. approximately 1% transgenesis. In fish whose eggs are protected by tough chorions (e.g. salmon and trout), it is often necessary to cut a hole prior to injection or exploit the micropyle

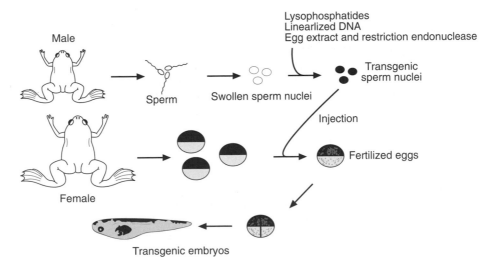

Figure 5.8

Procedure of Kroll and Amaya (6) for the production of transgenic *Xenopus*. Reprinted from Nature, **383**, Slack, JMW, *High hops of transgenic frogs*, p765, Copyright (1996), Nature Publishing Group.

for the introduction of DNA. More recently, methods such as electroporation and particle bombardment have also been utilized with success for fish transgenesis, as well as sperm-mediated gene transfer. A novel technique applied in some species is the electroporation of sperm with plasmid DNA followed by sperm-mediated DNA transfer, a procedure which is analogous in some ways to the REMI technique described for amphibians. This appears to be less efficient than injection, but has nevertheless resulted in transgenic carp, catfish, salmon and tilapia.

Like frogs, the injection of DNA into fish eggs and early embryos leads to extensive replication and expression from non-integrated transgenes, so that fish, like frogs, can be used for transient expression assays. Some of the DNA then integrates into the genome, leading to mosaicism, germline transmission, and the production of transgenic fish lines. There has been recent progress in the development of novel methods to enhance DNA integration in fish, based on the use of purified enzymes from integrating DNA elements (i.e. transposons and retroviruses). The successful transfer of exogenous genes to zebrafish embryos using retroviral vectors was followed by the development of a hybrid technique in which DNA was injected into zebrafish embryos along with purified retroviral integrase, resulting in earlier transgene integration and a significant increase in integration efficiency. This prompted a search for natural fish enzymes that could perform the same function, but unfortunately all the fish transposons that were isolated contained deletions or other mutations that prevented functional transposase expression. This problem was addressed by collecting the sequences of many fish transposons and deriving a theoretical ideal sequence, resulting in a synthetic transposon system known as *Sleeping Beauty*. The use of *Sleeping Beauty* as a gene transfer vector in zebrafish increases the efficiency of transgene integration over 20-fold, and could well be suitable for other species of fish. (See references (47,48).)

5.9 Transgenic insects and worms

The two major invertebrate model species are the fruit fly *Drosophila melanogaster* and the nematode *Caenorhabditis elegans*. In both cases, transformation is achieved by microinjection into the area of the embryo which will give rise to the germ cells – the pole plasm in the fly and the gonad in the worm. In the fly, gene transfer developed in the 1980s with the exploitation of endogenous transposons called P-elements, which could jump spontaneously into the genome of germplasm nuclei. In the worm, injection of plasmid DNA is the routine method for gene transfer, but the discovery of RNA interference in 1998 showed that small amounts of nucleic acids could be introduced into the worm simply by soaking in medium containing RNA, or feeding on bacteria expressing RNA. This is discussed in more detail in Chapter 9.

Gene transfer to *Drosophila*

The routine strategy used to introduce DNA into *Drosophila* embryos relies on P-elements, which are DNA transposons similar to the *Sleeping Beauty* elements discussed above. Unlike the *Sleeping Beauty* system, however,

P-elements are naturally occurring and fully functional transposons found in some strains of *Drosophila melanogaster* and related species. These transposons are only active in germ cells, but can be very active indeed, resulting in a phenomenon known as P–M hybrid dysgenesis where multiple genomic insertions result in a large number of mutations and chromosomal abnormalities in the offspring of certain crosses, causing a variety of genetic defects, sterility and abnormal development. Specifically, hybrid dysgenesis occurs when males of a strain containing P-elements (P type) are mated with females of a strain lacking P-elements (M type). It does not occur in crosses between M-type males and P-type females or in crosses where both the male and female are P-types. This reflects the fact that the P-type female produces a repressor of P-element transposition, thus transposition occurs only when P-elements in the sperm nucleus are introduced into an M-type egg (lacking the repressor).

The fully functional P-element is 2.9-kb in length with perfect 31-bp inverted terminal repeats, recognized by the transposase enzyme encoded within the body of the transposon. The transposase gene contains a number of exons, and through cell-specific differential splicing a functional product is produced only in germ cells. By manipulating the third intron of the transposase gene to prevent non-productive splicing, a derivative can be made which has the ability to transpose in somatic cells as well as germ cells. Not all the P-elements in the *Drosophila* genome are autonomous – in many cases the transposase gene is mutated or truncated and the shorter elements (sometimes described as P-factors) cannot mobilize under their own control. However, if the 31-bp repeats are intact, they can be mobilized in *trans* by transposase supplied by a functional P-element, and can likewise integrate into an endogenous gene and inactivate it by insertional mutagenesis.

The first procedure for *Drosophila* transformation was devised by Spradling and Rubin (7) and involved the injection of P-element DNA into the posterior pole of an M-type embryo at the syncytial blastoderm stage. As in other insects, *Drosophila* development involves a syncytial phase in which the zygote nucleus divides rapidly many times, but the daughter nuclei remain in a common cytoplasm (*Figure 5.9*). The injected construct was a plasmid containing a functional P-element in which some extra DNA (initially from the cloning vector pBR322) had been inserted between the 31-bp repeats. The posterior pole was chosen as the injection site because the nuclei in this region eventually form the germ cells, facilitating transformation of the germline. As expected, some of the progeny flies contained recombinant P-elements, which had transposed from the plasmid into random genomic sites. Although the use of a functional P-element for transposition facilitated efficient transformation of the *Drosophila* germline, a disadvantage of this early technique was the fact that further transposition events could occur in subsequent generations leading to transgene duplication and the mutation of endogenous genes.

A more sophisticated strategy was therefore developed in which an internally deleted P-element was used as the integrating construct. This element was incapable of autonomous mobilization, and could therefore jump from the plasmid to a genomic site only if supplied with transposase from another source. In an M-type *Drosophila* line, such a construct would thereafter

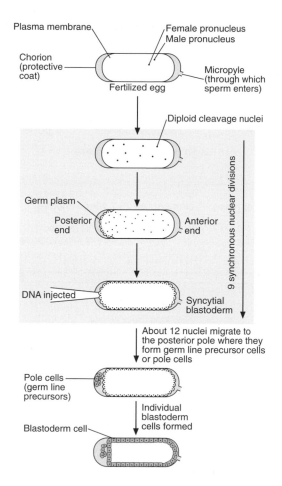

Figure 5.9

Early embryogenesis of *Drosophila*. DNA injected at the posterior end of the embryo just prior to pole cell formation is incorporated into germline cells. From Primrose, Twyman, Old (2001), Principles of Gene Manipulation, Blackwell Science.

remain immobilized at its initial integration site. Several strategies were developed for transformation with defective P-element vectors, including the co-injection of purified transposase and the use of dysgenic crosses where transposase is already available. However, the method that has stood the test of time is the binary system developed by Rubin and Spradling in 1983 (50,51) (*Figure 5.10*). This consists of a defective P-element vector in a standard cloning plasmid, consisting minimally of the transgene, a visible marker gene such as *white* or *rosy* (page 12) and the 31-bp inverted repeats. Integration of the vector into the fly genome is achieved using a helper element which is internally complete (i.e. able to supply functional transposase) but which has one or both of the inverted terminal repeats deleted so it cannot itself transpose – this is known as a wings-clipped element. The transposase from the wings-clipped element allows the defective vector to integrate into the genome, and the transgene is then fixed at the

Structure of P-element derivative: Carnegie 20

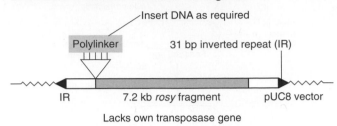

Lacks own transposase gene

Structure of helper P-element: pπ 25.7 wings clipped

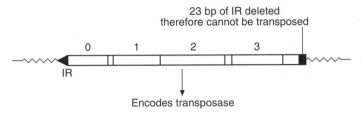

Encodes transposase

Figure 5.10

An early P-element-based transformation vector system for *Drosophila melanogaster*. See text for details. From Primrose, Twyman, Old (2001), Principles of Gene Manipulation, Blackwell Science.

integration site. The use of P-elements for gene transfer in *Drosophila* is considered again in Chapter 9, when we discuss large-scale insertional mutagenesis and gene trap programs. (See references (49–51).)

References

1. Jaenisch R, Mintz B (1974) Simian virus 40 DNA in DNA of healthy adult mice derived from preimplantation blastocysts injected with viral DNA. *Proc Natl Acad Sci USA* **71**: 1250–1254.
2. Palmiter RD, Brinster RL (1985) Transgenic mice. *Cell* **41**: 343–345.
3. Brinster RL, Chen HY, Trumauer M *et al.* (1981) Somatic expression of herpes thymidine kinase in mice following injection of a fusion gene into eggs. *Cell* **27**: 223–231.
4. Colman A (1984) Translation of eukaryotic messenger RNA in Xenopus oocytes. In: Hames BD, Higgens SJ (eds) *Transcription and Translation – A Practical Approach*, pp. 271–302. IRL Press, Oxford, UK.
5. Rusconi S, Schaffner W (1981) Transformation of frog embryos with a rabbit β-globin gene. *Proc Natl Acad Sci USA* **78**: 5051–5055.
6. Kroll KL, Amaya E (1996) Transgenic *Xenopus* embryos from sperm nuclear transplantations reveal FGF signaling requirements during gastrulation. *Development* **122**: 3173–3183.
7. Spradling AC, Rubin GM (1982) Transposition of cloned P elements into *Drosophila* germline chromosomes. *Science* **218**: 341–347.
8. Palmiter RD, Brinster RL (1986) Germline transformation of mice. *Annu Rev Genet* **20**: 465–499.

9. van der Putten H, Botteri FM, Miller AD, *et al.* (1985) Efficient insertion of genes into the mouse germline via retroviral vectors. *Proc Natl Acad Sci USA* **82**: 6148–6152.

10. Rubenstein JL, Nicolas JF, Jacob F, *et al.* (1986) Introduction of genes into preimplantation mouse embryos by use of a defective recombinant retrovirus. *Proc Natl Acad Sci USA* **83**: 366–368.

11. Evans MJ, Kaufman MH (1981) Establishment in culture of pluripotential cells from mouse embryos. *Nature* **292**: 154–156.

12. Martin GR (1981) Isolation of a pluripotent cell line from early mouse embryos cultured in medium conditioned by teratocarcinoma stem cells. *Proc Natl Acad Sci USA* **78**: 7634–7638.

13. Bradley A, Evans M, Kaufman MH, Robertson E (1984) Formation of germline chimeras from embryo-derived teratocarcinoma cells. *Nature* **309**: 255–256.

14. Pinkert CA, Murray JD (1999) Transgenic farm animals. In: Murray JD, Anderson GB, Oberbauer AM, McGloughlin MM (eds) *Transgenic Animals in Agriculture*, 1–18. CABI Publishing, Wallingford, UK.

15. Hammer RE, Pursel VG, Rexroad CE Jr, Wall RJ, Bolt DJ, Ebert KM, Palmiter RD, Brinster RL (1985) Production of transgenic rabbits, sheep and pigs by microinjection. *Nature* **315**: 680–683.

16. Love J, Gribbin C, Mather C, Sang H (1994) Transgenic birds by DNA microinjection. *Biotechnology* **12**: 60–63.

17. Bosselman RA, Hsu RY, Boggs T, Hu S, Bruszewski J, Ou S, Kozar L, Martin F, Green C, Jacobsen F (1989) Germline transmission of exogenous genes in the chicken. *Science* **243**: 533–535.

18. Chan AWS, Homan EJ, Ballou LU, Burns JC, Brennel RD (1998) Transgenic cattle produced by reverse transcribed gene transfer in oocytes. *Proc Natl Acad Sci USA* **95**: 14028–14033.

19. Chan AWS, Chong KY, Martinovich C, Simerly C, Shatten G (2001) Transgenic monkeys produced by retroviral gene transfer into mature oocytes. *Science* **291**: 309–323.

20. Lois C, Hong EJ, Pease S, Brown EJ, Baltimore D (2002) Germline transmission and tissue-specific expression of transgenes delivered by lentiviral vectors. *Science* **295**: 868–872.

21. Pfeifer A, Ikawa M, Dayn Y, Verma IM (2002) Transgenesis by lentiviral vectors: lack of gene silencing in mammalian embryonic stem cells and preimplantation embryos. *Proc Natl Acad Sci USA* **99**: 2140–2145.

22. Hofmann A, Kessler B, Ewerling S, *et al.* (2003) Efficient transgenesis in farm animals by lentiviral vectors. *EMBO Rep* **4**: 1054–1060.

23. Hofmann A, Zakhartchenko V, Weppert M, Sebald H, Wenigerkind H, Brem G, Wolf E, Pfeifer A (2004) Generation of transgenic cattle by lentiviral gene transfer into oocytes. *Biol Reprod* (in press).

24. Pfeifer A (2004) Lentiviral transgenesis. *Transgenic Res* (in press).

25. Pain B, Chenevier P, Samarut J (1999) Chicken embryonic stem cells and transgenic strategies. *Cells Tissues Organs* **165**: 212–219.

26. Kim JH, Junga HS, Lee HT, Chung KS, *et al.* (1997) Development of a positive method for male stem cell-mediated gene transfer in mouse and pig. *Mol Reprod Dev* **46**: 515–526.

27. Honaramooz A, Megee SO, Foley B, Borbrinski I, *et al.* (2003) Use of adeno-associated virus for transfection of male-germ cells for transplantation in pigs. *Theriogenology* **59**: 536.

28. Brinster RL (2002) Germline stem cell transplantation and transgenesis. *Science* **296**: 2174–2176.

29. Lavitrano M, Camaioni A, Fazio VM, Dolci S, *et al.* (1989) Sperm cells as vectors for introducing foreign DNA into eggs: genetic transformation of mice. *Cell* **57**: 717–723.

30. Lavitrano M, Forni M, Varzi V, *et al.* (1997) Sperm-mediated gene transfer: production of pigs transgenic for a human regulator of complement activation. *Transplant Proc* **29**: 3508–3509.

31. Lavitrano M, Bacci ML, Forni M, *et al.* (2002) Efficient production by sperm-mediated gene transfer of human decay accelerating factor (hDAF) transgenic pigs for xenotransplantation. *Proc Natl Acad Sci USA* **99**: 14230–14235.

32. Inoue K, Ogura A, Hayashi J-I (2002) Production of mitochondrial DNA transgenic mice using zygotes. *Methods* **26**: 358–363.

33. Moraes CT, Dey R, Barrientos A (2001) Transmitochondrial technology in animal cells. *Method Cell Biol* **65**: 397–412.

34. Briggs R, King TJ (1952) Transplantation of living nuclei from blastula cells into enucleated frog's eggs. *Proc Natl Acad Sci USA* **38**: 455–463.

35. Wilmut I, Schnieke AE, McWhir J, Kind AJ, Campbell KHS (1997) Viable offspring derived from fetal and adult mammalian cells. *Nature* **385**: 810–813.

36. Kato Y, Tani T, Sotomaru Y, Kurokawa K, Kato J, Doguchi H, Yasue H, Tsunoda Y (1998) Eight calves cloned from somatic cells of a single adult. *Science* **282**: 2095–2098.

37. Wakayama T, Perry AC, Zuccotti M, Johnson KR, Yanagimachi R (1998) Full-term development of mice from enucleated oocytes injected with cumulus cell nuclei. *Nature* **394**: 369–374.

38. Polejaeva IA, Chen SH, Vaught TD, *et al.* (2000) Cloned pigs produced by nuclear transfer from adult somatic cells. *Nature* **407**: 86–90.

39. Zou X, Chen Y, Wang Y, *et al.* (2001) Production of cloned goats from enucleated oocytes injected with cumulus cell nuclei or fused with cumulus cells. *Cloning* **3**: 31–37.

40. Shin T, Kraemer D, Pryor J, Liu L, Rugila J, Howe L, Buck S, Murphy K, Lyons L, Westhusin M (2002) A cat cloned by nuclear transplantation. *Nature* **415**: 859.

41. Chesne P, Adenot PG, Viglietta C, Baratte M, Boulanger L, Renard JP (2002) Cloned rabbits produced by nuclear transfer from adult somatic cells. *Nat Biotechnol* **20**: 366–369.

42. Galli C, Lagutina I, Crotti G, *et al.* (2003) A cloned horse born to its dam twin – A birth announcement calls for a rethink on the immunological demands of pregnancy. *Nature* **424**: 635.

43. Woods GL, White KL, Vanderwall DK, Li GP, Aston KI, Bunch TD, Meerdo LN, Pate BJ (2003) A mule cloned from fetal cells by nuclear transfer. *Science* **301**: 1063.

44. Schnieke AE, Kind AJ, Ritchie WA, Mycock K, Scott AR, Ritchie M, Wilmut I, Colman A, Campbell KH (1997) Human factor IX transgenic sheep produced by transfer of nuclei from transfected fetal fibroblasts. *Science* **278**: 2130–2133.

45. McCreath KJ, Howcroft J, Campbell KHS, Colman A, Schnieke AE, Kind AJ (2000) Production of gene targeted sheep by nuclear transfer from cultured somatic cells. *Nature* **405**: 1066–1069.

46. Slack JMW (1996) Developmental biology – High hops of transgenic frogs. *Nature* **383**: 765–766.

47. Iyengar A, Muller F, Maclean N (1996) Regulation and expression of transgenes in fish – a review. *Transgenic Res* **5**: 147–166.

48. Zbikowska HM (2003) Fish can be first – advances in fish transgenesis for commercial applications. *Transgenic Res* **12**(4): 379–389.

49. Ashburner M (1989) *Drosophila: A Laboratory Handbook*. Cold Spring Harbor Lboratory Press, New York.

50. Rubin GM, Spradling AC (1982) Genetic transformation of *Drosophila* with transposable element vectors. *Science* **218**: 348–353.

51. Rubin GM, Spradling AC (1983) Vectors for P-element mediated gene transfer in *Drosophila*. *Nucl Acids Res* **11**: 6341–6351.

Protocols

CONTENTS

MATERIALS

Avertin. The anesthetic Avertin is made by adding 10 g of tribromoethyl alcohol to 10 ml of tertiary amyl alcohol. This stock should be stored in the dark at 4°C. It should be diluted to 2.5% in sterile water to make the working solution.

70% ethanol

Dissection instruments

Silk sutures and clips

Bunsen burner

50 IU ml^{-1} follicle-stimulating hormone (made up in sterilized water and stored at -70°C)

50 IU ml^{-1} human chorionic gonadotrophin (made up in sterilized water and stored at -70°C)

M2 and M16 egg culture medium. These are made up from concentrated stocks A, B, C, D and E as described in *Table 5.2.*

Liquid paraffin

10 mg ml^{-1} hyaluronidase in M2

Egg transfer pipettes (these can be purchased, but acceptable pipettes can be made by stretching a standard BDH glass capillary in a Bunsen flame, scoring the glass with a diamond pencil and snapping where the tube is about 250 μm across).

DNA for microinjection, linearized and prepared by column chromatography

Microinjection apparatus, comprising inverted dissecting microscope, two micromanipulators, micrometer syringe linked to a holding pipette via plastic tubing filled with paraffin, Pico-injector linked to the injection pipette via air-filled plastic tubing, compressed air supply. The holding and injection pipettes can be purchased, but again they can easily be made. Holding pipettes are made from hard glass capillaries stretched in a Bunsen flame, finished with a diamond pencil and a microforge. Injection pipettes need to be about 1 μm wide at the tip, and are made using a mechanical pipette puller

Table 5.2   Ingredients for M2 and M16 egg culture medium.

	COMPONENT	g/100 ml
STOCK A (10×)	NaCl	5.534
	KCl	0.356
	KH_2PO_4	0.162
	$MgSO_4 \times 7\ H_2O$	0.293
	Sodium lactate	2.610 or 4.349 g of 60% syrup
	Glucose	1.000
	Penicillin G	0.060
	Streptomycin	0.050
STOCK B (10×)	$NaHCO_3$	2.101
	Phenol Red	0.001
STOCK E (10×)	HEPES	5.958
	Phenol Red	0.001

	COMPONENT	g/10 ml
STOCK C (100×)	Sodium pyruvate	0.036
STOCK D (100×)	$CaCl_2 \times 2\ H_2O$	0.252

STOCK	M2 (ml)	M16 (ml)
A (10×)	10.0	10.0
B (10×)	1.6	10.0
C (100×)	1.0	1.0
D (100×)	1.0	1.0
E (10×)	8.4	–
WATER	78.0	78.0
BSA (bovine serum albumin, Sigma A4378	400 mg	400 mg

Protocol 5.1: Preparation of vasectomized male mice

1. Anesthetize a healthy, 2-month-old male mouse with Avertin made up as shown on page 127, using 15 µl of the working solution per gram body weight. Restrain the mouse by holding the back pelt firmly in one hand, exposing the belly. Inject into the abdomen using a 0.5 × 16 mm needle, making sure to avoid the bladder and diaphragm. The mouse should be unconscious for at least 30 minutes.

2. Place the unconscious mouse with the abdomen uppermost and spray the abdominal area with 70% ethanol. Using a pair of forceps, lift the skin at the incision site, which is level with the top of the hindlimbs, and make a cut in the body wall. Immediately place a stitch through the body wall adjacent to the incision.

3. Pull one testis and the associated fat pad through the opening with a pair of fine forceps, exposing the vas deferens. Free the vas deferens from its membranes.

4. Hold the vas deferens with one pair of forceps and twist through 180° to make a loop. Heat a pair of blunt forceps in the Bunsen flame until red hot. Then use the heated forceps to grip the vas deferens at the loop. This will burn through and cauterize the cut ends.

5. Replace the testis and fat pad inside the body. Then repeat the procedure for the other vas deferens.

6. When the second testis has been replaced, sew up the body wall and clip the edges of the incision together.

7. Vasectomized males should be placed in individual cages and left for at least 3 weeks before mating.

Protocol 5.2: Superovulation of immature females

1. In the morning of day 1, take a healthy female mouse about 1 month old and inject five units of follicle-stimulating hormone (0.1 ml of stock solution) intraperitoneally. Replace the mouse in the cage and leave for 2 days.

2. On the morning of day 3, perform the same procedure with five units of human chorionic gonadotrophin (0.1 ml of stock solution). Place each female with a stud male and leave overnight.

3. On the morning of day 4, check the females for the presence of a copulatory plug.

Protocol 5.3: Collection of fertilized eggs

1. Prepare a number of culture dishes or one six-well plate containing 2–3 ml per well of M16 egg culture medium. Also prepare culture dishes in which drops of M16 medium are covered in paraffin. These should be placed in an incubator at 37°C and 5% CO_2.

2. Prepare one six-well plate containing 2–3 ml per well of M2 egg culture medium and store at room temperature under a laminar hood.

3. Sacrifice the donor female by cervical dislocation and spray the abdominal area with 70% ethanol.

4. Make an incision in the abdomen wall and withdraw the oviduct, using a second pair of forceps to trim away the mesometrium attaching the reproductive tract to the inner body wall.

5. Make two cuts to separate the oviduct from the uterus and from the ovary and place the entire ovary into one of the dishes containing M2 medium. Repeat this process on the other side of the animal, and on the other donor females.

6. When all the oviducts are dissected out, transfer them one at a time to fresh medium and place the well under a dissecting microscope. Carefully tear the ampulla to release the mass of cumulus cells and the fertilized eggs. Discard the remains of the oviduct and repeat for the other oviducts.

7. Add 50 ml of hyaluronidase stock and mix to help release the eggs. After 2–3 minutes, transfer the eggs to fresh M2 medium, trying to leave behind as much of the cumulus mass as possible. Pipetting up and down a few times may help. Repeat this process and transfer to another fresh dish of M2 medium.

8. Transfer the eggs to the prewarmed and pregassed dishes of M16 medium and wash.

9. Transfer the washed eggs to the M16 droplets under paraffin and incubate at 37°C in 5% CO_2 until needed.

Protocol 5.4: Pronuclear microinjection

1. The microinjection apparatus should be set up under a laminar flow hood. For a right-handed operator, the holding pipette is controlled by the left micromanipulator and is connected to the micrometer syringe, while the injection pipette is controlled by the right micromanipulator and is connected to the Pico-injector.

2. Remove 20–30 eggs from the droplets of M16 medium and wash briefly in M2 medium, and place in the injection chamber in a small droplet of medium avoiding air bubbles.

3. With a single egg in sharp focus, adjust the vertical positions of the two pipettes so that the pipettes and egg are in the same plane.

4. Apply gentle suction to the holding pipette by withdrawing the syringe barrel very slightly. Position the egg centrally in the field of view. Focus up and down until the (larger) male pronucleus is visible. Use the injection needle and variable suction on the holding pipette to rotate the egg, if necessary, to position the male pronucleus towards the injection needle.

5. Once everything is in line, push the needle into the egg and inject the DNA into the nucleus. A successful injection results in the nucleus swelling to about twice its normal size. If nothing happens, the needle may be blocked and may need to be changed.

6. Withdraw the needle and use the holding pipette to move the injected egg to one side. Continue injecting the other eggs and collect the injected ones together, separate from the uninjected ones. The eggs should not leak significantly after injection. If the cytoplasm appears to be leaking out, the needle should be changed.

7. Leave eggs in M2 medium for a maximum of 15 minutes, then transfer the batch of injected eggs to M16 medium for washing, and then back under the droplets under paraffin.

Protocol 5.5: Transfer of injected eggs into foster females

1. Place an ovulating female with one of the vasectomized males produced in *Protocol 5.1*, in the afternoon before the day of egg transfer.

2. The next morning, check for a copulatory plug. Anesthetize pseudopregnant females as described in *Protocol 5.1*.

3. Place the unconscious mouse face down on the lid of a Petri dish and spray the back with 70% ethanol.

4. Make a transverse incision in the skin about 1 cm long, level with the last rib. Then make a smaller incision in the body wall adjacent to the ovary, which can be seen underneath. Place a stitch in the body wall ready for closure.

5. Pull on the fat pad which is joined to the ovary, and the reproductive system will follow through the incision. Clip the fat pad so that the oviduct remains uppermost. Place the mouse under a dissecting microscope.

6. Carefully tease open the transparent membrane covering the ovary and identify the infundibulum, which is the opening to the oviduct. The tear should be made as close to the infundibulum as possible.

7. When the recipient is ready, remove 10–15 eggs from the droplet of M16 medium and wash gently in M2 medium. Take up the eggs in a transfer pipette, but first fill it almost completely with paraffin oil, then take up a small amount of M2 medium separated from the paraffin by an air bubble. Finally take up the eggs, separated from the M2 medium by another small air bubble. Allow some more air into the pipette after the eggs.

8. With the transfer pipette on standby, grip the infundibulum with a sharp pair of forceps and bring it through the tear in the ovary membrane. Inject the eggs through the opening, watching for the appearance of three air bubbles to confirm that the eggs have been deposited.

9. Withdraw the pipette, unclip the fat pad and replace the reproductive tract inside the abdominal cavity. Sew the body wall and clip the skin together at the site of the incision.

10. If required, the procedure can be repeated on the other horn of the reproductive tract.

11. Transfer the foster animal to an individual cage for recovery. If successful, pups should be born 20 days after the operation.

Protocol 5.5: Transfer of injected eggs into foster females

Gene targeting in animal cells

6

6.1 Introduction

In the previous four chapters, we have discussed procedures for the genetic manipulation of animal cells and whole animals, always involving the *addition* of genetic information to the host cell or organism. The new DNA sequences may be maintained episomally in cell lines if the vector contains an origin of replication that is functional in animal cells. Alternatively, stable transformation may be achieved by integration of the exogenous DNA into the host genome, an approach that is always required for the generation of transgenic animals. The mechanism of integration in most cases is a process known as *illegitimate recombination*, which is based on the end-joining of non-homologous DNA fragments and involves a ubiquitous set of proteins with no particular sequence requirements. In a few cases, such as the transformation of *Drosophila* with P-elements and the integration of retroviral/lentiviral vectors, a slightly different mechanism known as *transpositional recombination* is involved. Here, the proteins mediating the integration process (enzymes known as transposases or integrases) recognize specific sequences on one of the recombining partners (the transposon or retroviral genome carrying the foreign DNA) while the DNA sequence at the recipient site in the genome is relatively non-specific. The result in either case is that DNA integration into the host genome is a more-or-less random process.

While such approaches are extremely useful, it is often desirable to carry out more precise manipulations in which specific endogenous sequences are modified or replaced with exogenous DNA. This allows genes to be targeted for disruption, deletion, rearrangement or replacement and facilitates the introduction of precise subtle mutations into preselected sequences. The general term used to describe such operations is *gene targeting*. Precise modifications cannot be achieved by either illegitimate or transpositional integration mechanisms because there is no control over the transgene integration site. Gene targeting requires alternative forms of recombination in which both the donor and recipient DNA sequences are defined, and two such recombination mechanisms have been identified. The first is *homologous recombination*, a form of genetic exchange which occurs in all cells, and requires two DNA molecules with similar, but not necessarily identical, sequences. No specific sequences are required but, at least in animal cells, long regions of homology are necessary for efficient exchange. The frequency of homologous recombination in animal cells is generally very low compared to illegitimate recombination, and sophisticated selection strategies are usually required to identify cells in which the exogenous DNA has recombined with the endogenous genome rather than integrating randomly. The second is *site-specific recombination*, which involves proteins that recognize short and very specific recombinogenic sequences. Unlike homologous recombination, which is a ubiquitous process, different organisms have their own site-specific recombination systems, meaning that heterologous systems can be imported into animal cells and used to control transgene behavior precisely. Some of the most sophisticated strategies for controlling the structure and activity of transgenes and endogenous genes rely on imaginative combinations of homologous and site-specific recombination. These combined systems are discussed at the end of the chapter, including the recent development of a robust gene-targeting system for the fruit fly *Drosophila melanogaster*. The properties of different recombination mechanisms are summarized in *Table 6.1*.

6.2 History of gene targeting by homologous recombination in mammalian cells

Gene targeting by homologous recombination is very efficient in bacteria and yeast because these organisms have small genomes and homologous recombination is a major DNA repair mechanism for broken DNA strands. However, when the first gene-targeting experiments were carried out in animal cells in the 1980s, only a very low frequency of targeted recombination events was achieved. It is thought that this reflects the larger genome size in animal cells, and the fact that non-homologous end-joining is the major repair mechanism for broken DNA strands. These initial experiments involved the correction of mutations in selectable markers such as *neo*, which had been introduced into cell lines as transgenes by standard

Table 6.1 Different forms of recombination exploited for the genetic manipulation of animal cells

Mechanism	Uniqueness	Donor site	Recipient site	Major uses
Illegitimate	Ubiquitous	Non-specific	Non-specific	Transgene integration
Transpositional	Different systems encoded by different transposons and retroelements	Specific (usually inverted or direct repeats)	Non-specific but in many cases a weak consensus at the recipient site may be required	P-element integration, retroviral integration, insertional mutagenesis
Homologous	Ubiquitous	Non-specific but must be homologous to recipient site	Non-specific but must be homologous to donor site	Gene targeting, transgene insertion into some viral vectors
Site-specific	Different systems in different organisms	Specific	Specific	Site-specific transgene integration and excision, conditional (in)activation
In vitro	Uses purified DNA ligase	Blunt or compatible sticky ends	Blunt or compatible sticky ends	Construction of recombinant DNA molecules

methods. Smithies *et al.* (1) were the first to demonstrate the targeting of an endogenous gene. They introduced a modified β-globin gene containing the bacterial marker *supF* into a human fibroblast x mouse erythroleukemia cell line and screened large numbers of cells by recloning the marker. This experiment demonstrated that the frequency of homologous recombination was 10^3–10^5 times lower than that of random integration.

As discussed below, however, gene targeting is much more efficient in a small number of specific cell lines, one of which happens to be murine ES cells. This allows ES cells to be used not only for the production of transgenic mice carrying extra DNA sequences (see Chapter 5) but also gene-targeted mice with the same predefined mutation in every cell, including the germline. The first report of successful gene targeting in ES cells involved the disruption of the *Hprt* locus with the selectable marker gene *neo*, allowing targeted cells to be selected using the antibiotic G418 (2). The *Hprt* gene was chosen because it is found on the X-chromosome, and since most ES cell lines are male, only one gene-targeting event was required to eliminate the locus. (See references (1,2).)

6.3 Target cells for homologous recombination

In theory, any mammalian cell can be targeted by homologous recombination but the targeting efficiency in most cells is very low. The vast majority of experiments published to date have involved murine ES cells and, until recently, mice were the only animals that could be genetically manipulated by homologous recombination. ES cells can be transfected by any of the methods described in Chapter 2, although electroporation and lipofection have provided the highest transfection efficiencies and have proven to be the most compatible with germline transmission (*Protocol 6.1*). Lipofection is particularly suitable for the introduction of large DNA fragments, such as YACs and the newer generation of BAC-based targeting vectors, although for very large chromosome fragments the technique of microcell-mediated gene transfer has been successful (Chapter 3).

Recently, there has also been considerable interest in the use of somatic cells for gene targeting due to the development of nuclear transfer technology in domestic mammals. In most somatic cells, the efficiency of gene targeting is lower than 10^{-4}, so thousands or tens of thousands of clones need to be screened to identify a single targeting event. For established cell lines, such constraints may be acceptable, but since primary cells are usually isolated in only small numbers, and are able to divide only a few times in culture prior to senescence, the use of such cells for gene targeting is limited. In attempts to overcome this limitation, various enrichment strategies have been developed, including the use of positive–negative selection and promoterless vector systems (see below). A small number of somatic cell lines have been identified which, like ES cells, appear to be particularly amenable to gene targeting. Once example is the chicken pre-B cell line DT40, although the reasons for the high targeting efficiency remain unclear. No equivalent mammalian cells have been identified, but DT40 cells can be used to direct the genetic modification of mammalian cells in a surrogate environment. For example, by combining gene targeting in

DT40 cells and microcell fusion, specific mutations can be introduced into human or rodent chromosomes that have been transferred into DT40 host cells, and the chromosomes can then be shuttled back into a mammalian cell line for functional analysis. (See references (3–5).)

6.4 Design of targeting vectors

Gene targeting by homologous recombination can be used to achieve various goals (*Table 6.2*) but in all cases the vectors used to introduce the exogenous DNA have a number of common properties. Targeting vectors are specialized plasmid vectors that promote homologous recombination when introduced into ES cells and other cell types. This is achieved by the inclusion of a *homology region*, i.e. a region that is homologous to the target gene, allowing the targeting vector to pair up with the endogenous DNA sequence. The size of the homology region and the level of sequence identity have been shown to play an important role in the efficiency of gene targeting, with 2 kb of homology as the minimal requirement in mammalian cells. The most recent generation of targeting vectors is based

Table 6.2 Gene targeting by homologous recombination can be used to achieve different goals.

Targeting strategy	Explanation
Gene knockout	Disruption or deletion of a targeted endogenous gene. Disruption can be achieved by the targeted insertion of a marker gene such as *neo* in a single step, while deletion requires two sequential targeting events. In many cases, it is not necessary to delete the entire gene – the removal of a single exon may be sufficient.
Allele replacement	The introduction of a replacement allele in a targeted endogenous gene, usually distinguished from the above in that the replacement allele is a subtle variation (e.g. a point mutation) rather than a null allele. This always requires a two-step procedure, either two sequential rounds of homologous recombination, or homologous recombination followed by site-specific recombination.
Gene repair	A form of allele replacement where a dysfunctional mutant allele is replaced by a functional allele. Requires two sequential targeting events.
Targeted insertion	A method for transgene insertion at a specific locus, without necessarily disrupting any endogenous genes. Requires a single targeting event.
Gene knock-in	The replacement of an endogenous gene with another *non-allelic* sequence. Used, for example, to express one gene using the full complement of endogenous regulatory elements, in the correct genomic context, of another.
Gene knock-down	A term included in this table for completeness only, which refers to any strategy for reducing the expression of an endogenous gene. May or may not be based on gene targeting.

on bacterial artificial chromosomes and can carry over 50 kb of flanking sequence. In the best cases, this has increased the efficiency of homologous recombination to 1.5×10^{-1} compared to non-homologous integration events. The level of sequence identify is more important when the amount of flanking sequence is low. Te Riele and colleagues (8) noted a 20-fold increase in targeting efficiency when the targeting construct was prepared using DNA that was isogenic (genetically identical) to the recipient ES cells, but this constraint is at least partially alleviated through the use of BAC vectors with extensive homology regions. Recombination is also more efficient if the vector is linearized prior to transfection.

Most gene-targeting experiments have been used to disrupt endogenous genes by inserting a large cassette within an essential exon, resulting in targeted null alleles (a strategy often termed *gene knockout*; see *Table 6.2*). Two types of targeting vector have been developed for this purpose: insertion vectors and replacement (or transplacement) vectors. Insertion vectors (*Figure 6.1*) are linearized *within* the homology region, and this causes the entire vector to integrate into the target locus. Insertion not only disrupts the target gene but also leads to a duplication of the sequences in the homology region. This can be problematical because the duplicated sequences can undergo a further recombination event, restoring the endogenous locus to its original state. Replacement vectors (*Figure 6.2*) are linearized *outside* the homology region, such that the entire homology region is collinear with the target. Targeting is achieved by two crossover events within the homology region, and only sequences within the homology region are inserted. Therefore, in order to achieve gene knockout, the homology region itself must be interrupted with a cassette.

Although both types of vector are designed to generate a null phenotype, the experiment must be designed carefully to ensure that an essential region

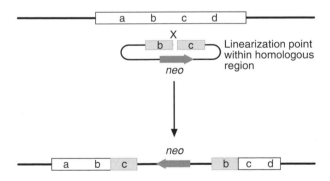

Figure 6.1

Insertion vector method. The introduced vector DNA is cut at a unique site within a sequence which is identical or closely related to part of the target endogenous gene. Homologous recombination (X) can occur, inserting the entire vector sequence (including the marker gene *neo*, which confers resistance to the antibiotic G418) into the targeted locus. However, this selection strategy does not allow genuine targeting events to be distinguished from random integration and further analysis (e.g. by PCR) is required. Taken from Strachan *et al, Human Molecular Genetics*, Copyright (2004), Garland Science.

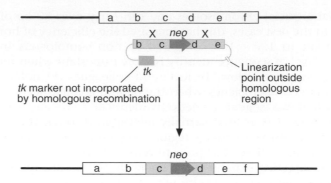

Figure 6.2

Replacement vector method. In this case, the *neo* gene is contained within the sequence homologous to the endogenous gene, and the vector is cut at a unique location outside the homology region. A double recombination or gene conversion event (X X) can result in replacement of internal sequences within the target gene by homologous sequences from the vector, including *neo*. Note that with replacement vectors it is possible to use a counterselectable marker gene such as *Tk*, placed outside the homology region to select against random integration events. *Tk* will not be inserted into the locus by homologous recombination, so cells that are resistant to G418 and sensitive to ganciclovir are genuine targeting events. Taken from Strachan *et al*, *Human Molecular Genetics*, Copyright (2004), Garland Science.

of the gene is targeted. There have been a number of reports in which gene targeting was successful, but the targeted gene continued to be expressed because transcription through the cassette allowed a functional gene product to be generated by alternative splicing. (See references (6–9).)

6.5 Selection strategies

The first gene-targeting experiments involved a locus (*Hprt*) whose disruption resulted in an easily selectable phenotype. However, most experiments do not involve such convenient loci and a selectable marker gene such as *neo* is often introduced as part of the targeting vector to allow the identification and selective propagation of transformed cells. Selectable markers are not always required – instead, large numbers of cells can be screened by PCR to identify genuine targeting events. However, a number of useful strategies involving selectable markers have been devised, and are discussed below.

When insertion vectors are used with selectable markers, the position of the marker is unimportant because the entire vector backbone integrates into the genome. In replacement vectors, it is important to include the marker within the homology region, as only this part of the vector integrates. Indeed, the selectable marker is generally used as the disruption cassette. The *neo* marker has been used most frequently, allowing transformed cells to be selected using G418. This strategy is shown in *Figure 6.1*. However, a single marker fails to discriminate between targeted cells and those where the construct has integrated randomly. Where replacement vectors are used, this issue can be addressed by a strategy known as positive–negative selection using *neo* as the positive marker and the HSV *Tk* gene as a negative or counterselectable marker. The *neo* marker interrupts the homology region as in standard vectors, allowing transformed cells to be selected using G418. However, the *Tk* gene is sited outside the homology region, so it does not integrate by homologous recombination but *does* integrate by illegitimate recombination. Therefore, cells that have undergone homologous recombination will survive in the presence of toxic thymidine analogs such as ganciclovir or FIAU (1-(2-deoxy-2-fluoro-β-D-arabinofuranosyl)-5-iodouracil), which are incorporated into DNA resulting in cell death only if there is a functional *Tk* gene expressed in the cell. This strategy is shown in *Figure 6.2*.

A further strategy is to make expression of the *neo* gene dependent on homologous recombination using a so-called *promoterless neo vector*. In this type of construct, the *neo* gene has no intrinsic promoter and therefore is not expressed following random integration. However, the marker comes under the control of the endogenous gene's promoter following homologous recombination, allowing targeted cells to be selected. A disadvantage of this method is that it only works with genes that are expressed in ES cells. This can be addressed by expressing the marker under the control of its own (constitutive) promoter, but making the cassette dependent on the target gene for polyadenylation (page 212). (See references (10,11).)

6.6 Introducing point mutations by gene targeting

The gene-targeting strategies discussed above result in a targeted locus containing a large, disruptive cassette containing a marker gene often under the control of its own regulatory elements, which can disrupt the expression of surrounding genes. In order to investigate the effects of more subtle mutations, such as small deletions or point mutations, more sophisticated approaches are required. Two major strategies for the introduction of subtle mutations have been developed, each requiring two rounds of homologous recombination.

Where insertion vectors are preferred, the 'hit and run' or 'in–out' strategy can be used (*Figure 6.3*). Only one insertion vector is required, but it must contain the desired subtle mutation within the homology region, and positive and negative markers such as *neo* and *Tk* somewhere on the backbone. The initial targeting event inserts the markers into the targeted locus and results in the duplication of the homology region (one copy of which carries the desired subtle mutation). This event can be positively selected by adding G418 to the medium, although cells in which the vector has integrated randomly will also be selected at this stage. If targeting has occurred, a second recombination event may then take place in which there is a crossover between the duplicated homology regions. In 50% of cases, this will eliminate the markers and restore the original sequence, while in the remaining cases, this will eliminate the markers and replace the original sequence with the subtle mutation. These second round recombinants will survive in the presence of ganciclovir because the *Tk* marker is eliminated. The surviving cell clones must be screened by PCR to identify wild types and targeted mutants.

The 'tag and exchange' strategy is used with replacement vectors (*Figure 6.4*). This not only requires two consecutive homologous recombination

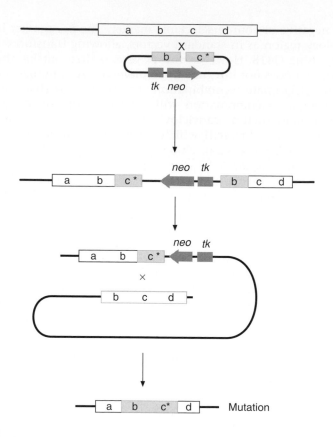

Figure 6.3

In the 'hit and run' strategy, used with insertion vectors, the subtle mutation is present on the first targeting construct, which contains both a selectable marker (*neo*) and a counterselectable marker (*tk*). Intrachromosomal recombination leads to the elimination of the marker gene and vector backbone. Taken from Strachan *et al*, *Human Molecular Genetics*, Copyright (2004), Garland Science.

events, but also two separate vectors. The example shown in *Figure 6.4* again utilizes the *neo* and *Tk* markers. The first 'tag' vector contains both markers within the homology region, and is designed to mutate the target gene by inserting the markers as a large cassette. In contrast to the hit-and-run strategy, the first targeting vector does *not* carry the subtle mutation. Targeting events (and random integrations) are selected with G418. The second 'exchange' vector has an intact homology region (no markers) but carries the desired mutation. Homologous recombination replaces the original event (the tag) with the sequence carrying the subtle mutation. This event can be selected with ganciclovir, which is non-toxic in the absence of *Tk*. (See references (10,11).)

6.7   Gene targeting in mice

In Chapter 5, we discussed the various methods available for the production of transgenic mice, one of which was the transfection of ES cells and

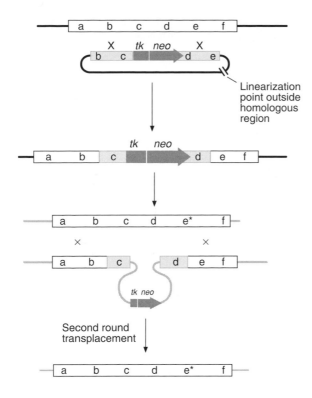

Figure 6.4

In the 'tag and exchange' strategy, used with replacement vectors, a second targeting construct containing the subtle mutation is used to replace the cassette, containing the selectable and counterselectable markers, introduced by the first round of targetting. First-round replacement is selected by G418 resistance (presence of *neo*), and second-round replacement is selected by ganciclovir resistance (absence of *tk*). Taken from Strachan *et al*, *Human Molecular Genetics*, Copyright (2004), Garland Science.

their subsequent incorporation into host blastocysts, resulting in chimeric animals in many cases showing germline transmission of the transgene. By combining this method with gene targeting in ES cells, it is possible to produce not only mice carrying additional genetic information, but also mice with specific mutations introduced into preselected endogenous genes. The general procedure is shown in *Figure 6.5*, and is simply an extension of the technique described in Chapter 5 (*Figure 5.3*). In the example shown, the aim is to create a deletion which eliminates a particular endogenous mouse gene, resulting in a so-called *gene knockout mouse*. The deletion event is created by gene targeting in a population of ES cells using one of the methods discussed above, selecting and validating the desired targeting events in culture (*Protocol 6.2*) and then introducing targeted ES cells into recipient blastocysts (*Protocol 6.3*). As discussed in Chapter 5, most ES cell lines are male and are derived from mouse strain 129, which has a dominant agouti coat color. These ES cells are introduced into blastocysts from a strain such as C57B10/J, which has a recessive coat color marker. The blastocysts are implanted into pseudopregnant females and brought to term,

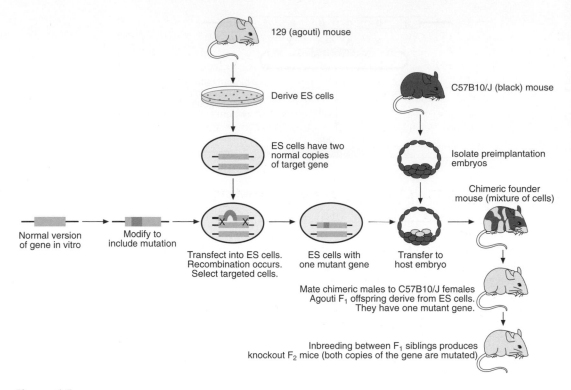

Figure 6.5

Procedure for generating homozygous gene knockout mice by gene targeting in ES cells. See text for details.

the resulting mice being cell chimeras (i.e. containing sectors of tissue derived from ES cells and the host inner cell mass, as identified by the patchwork of coat colors). Male chimeras are then mated with C57B10/J females, producing offspring with either agouti or black coats. Those with the agouti coat are derived entirely from germline tissue carrying the genetic modification. Importantly, if the targeted locus lies anywhere other than the X- or Y-chromosome, then first-generation siblings need to be mated together to produce mice that are homozygous for the mutation. The zygosity of the second-generation mice can be established by Southern blot hybridization of DNA taken from tail tips, as shown in *Figure 6.6*.

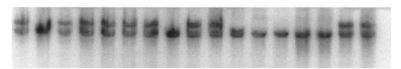

Figure 6.6

Genomic Southern blot of DNA from tail tips of F2 mouse litter from F1 sibling intercross, showing homozygous (single band) and heterozygous (two bands) individuals. Image courtesy of Dr John O'Brien, Neurobiology Division, MRC Laboratory of Molecular Biology, Cambridge, UK.

Coat color markers in gene-targeted mice can also be put to other uses. For example, Zheng *et al.* (17) used a tyrosinase minigene within their targeting construct to track the targeted allele in albino mice. Tyrosinase is required for melanin biosynthesis, and its expression leads to pigment production and thereby to a detectable gray coat color in an albino background. An agouti minigene has been used in a similar manner, to darken the yellow color in a wild-type agouti background. The depth of color can also help to distinguish heterozygotes and homozygotes in the F2 generation, although molecular analysis is required for confirmation. (See references (14–17).)

6.8   Enhancing the efficiency of gene targeting

An early observation in the history of gene targeting was that linear vectors are more effective than circular ones, suggesting that homologous recombination is facilitated by the presence of free DNA ends. Perhaps partly for this reason, recently developed targeting vectors based on adeno-associated virus (Chapter 4) have been shown to be highly efficient for gene targeting in mouse cells. Using such vectors, the targeting construct is presented as a linear, single-stranded DNA molecule, which facilitates gene targeting by invading the homologous duplex. The presence of double-stranded breaks enhances this process even further.

The existence of obscure cell lines such as DT40, which show an unusually high targeting efficiency, indicates that it may be possible to manipulate other cell lines, either genetically or through the direct activation or inhibition of gene products, to likewise enhance the efficiency of gene targeting. Genes involved in the processes of homologous recombination and illegitimate recombination are among the most promising targets, and there is a large body of evidence showing that the mutation or inhibition of certain key proteins can reduce the efficiency of gene targeting, perhaps suggesting that overexpression strategies might have the opposite, beneficial effect. For example, mutations affecting the two Rad54 homologs in ES cells reduce gene-targeting efficiency, and the normal function of these proteins is to promote DNA strand exchange mediated by another component of the homologous recombination machinery, Rad51. Another protein that interacts with Rad51 is Brca1, the tumor suppressor protein linked to breast cancer in humans. Again, mutation of the gene encoding Brca1 in ES cells reduces the efficiency of gene targeting and homologous recombination generally. The impact of p53 activity on gene targeting in ES cells has also been investigated. Perhaps the most promising experiments thus far have involved deliberate manipulation of Rad51 itself. The human *RAD51* gene has been overexpressed in human cells and improved the frequency of gene targeting up to three-fold, and similar results have been reported for the murine *Rad51* gene in ES cells.

There have been no reports thus far involving the manipulation of proteins controlling illegitimate recombination, although it is likely that inhibiting these proteins may favor homologous recombination. However, mutations in mismatch repair genes do appear to reduce the efficiency of targeting, probably by suppressing recombination when the sequences of the target and vector diverge.

A direct treatment which can improve the gene-targeting efficiency in cultured cells is the addition of 1,5-isoquinolinediol to the culture medium. This reagent inhibits the enzyme poly(ADP-ribose) polymerase (PARP), and results in a near ten-fold increase in targeting events. For an unknown reason, however, this approach only works when the targeting vector is chemically transfected into the cells, not when electroporation is used as the gene transfer method. (See references (18–21).)

6.9   Gene targeting in domestic mammals

With the lack of ES cells for most domestic species, gene targeting in mammals other than mice has been achieved in somatic cells followed by nuclear transfer to produce genetically modified animals. Such targeting events are extremely rare and, at time of publication, has only been achieved using the promoterless-type vectors described in Section 6.5, which must insert within the gene in such a way that the selectable marker can be expressed from the endogenous promoter. In the first such report, McCreath and colleagues (25) inserted the human gene for α1-antitrypsin into the 3′ untranslated region of the sheep α1(I) procollagen (*COL1A1*) locus. The construct was designed so that the human protein, which is used to treat cystic fibrosis and similar diseases, was expressed in the mammary gland and secreted into the milk. This gene-targeting strategy involved site-specific insertion of a transgene rather than deletion of endogenous coding sequence and hence comes under the heading of targeted insertion rather than gene knockout (*Table 6.2*). The first gene knockouts in domestic mammals were reported in 2001, and resulted in the production of fetuses and live lambs with targeted disruptions in either the *PRP* gene, which encodes the prion protein, or the *GGTA1* gene, which encodes the enzyme α(1,3)-galactosyl transferase. The *PRP*-knockouts represented the first domestic mammals to be generated by gene targeting specifically for the purpose of disease resistance, while the purpose of the *GGTA1* knockouts was to remove an important epitope found on the surface of all mammalian cells except those of apes and humans, which is a major cause of hyperacute rejection in xenotransplants (page 237). In 2002, two groups independently reported targeted disruption of the *GGTA1* gene in pigs, whose organs are similar in size to their human counterparts and are envisaged as the most likely source for xenotransplants (27,28). In each of the reports cited above, the target gene was autosomal and only one allele was disrupted. The first homozygous *GGTA1* knockout pigs were produced by Phelps *et al.* in 2003 (29) using heterozygous cells derived from the earlier cloned pigs (28). The aim of the experiment was to disrupt the second *GGTA1* allele by gene targeting and then select cells lacking the gal-epitope using a fungal toxin. However, on further analysis it was established that the second allele had been produced by a spontaneous point mutation within the *GGTA1* gene, so in the strict sense the pigs were still heterozygous, although for two null alleles.

One fact that is clear from these experiments is that gene knockout in domestic animals via gene targeting in somatic cells followed by nuclear transfer combines two already-inefficient processes and thus requires many unsuccessful attempts before a live animal is born. There is a funnel effect resulting from poor efficiency at each stage of the procedure: primary cells have a limited lifespan, so most transfected colonies undergo senescence before they can be expanded sufficiently for use in nuclear transfer, and then a large proportion of the transfers themselves also fail. In order to improve the efficiency of gene targeting in livestock, methods are needed that prolong the lifetime of the cultured cells during the gene-targeting process, allow gene targeting to be achieved with fewer cell divisions and make it easier to reprogram the somatic nucleus following transfer to the egg. Cells go through 40–50 divisions during the process of gene targeting (including drug selection), which is sufficient to bring many cells close to senescence. The lifespan of, for example, a sheep fetal fibroblast is about 100 divisions, but explants used to derive somatic cells for gene targeting are always heterogeneous populations with some cells having undergone more divisions than others. In human cells, it appears that there is some random aspect to longevity and that cells have a 'half-life' of 8–10 divisions. Cells of livestock mammals are likely to behave in a similar manner, so the number of cells with the potential for successful nuclear transfer probably falls with every division, providing some explanation for the relatively low efficiency of cloning. Very little is known about the mechanism of nuclear reprogramming, although there have been studies in *Xenopus*, mammals and yeast indicating a role for various chromatin remodeling proteins and transcription factors. In the future, it may therefore be possible to 'prime' somatic cells with the appropriate proteins to expedite nuclear reprogramming when the somatic nucleus is introduced into the egg. A final important consideration is telomere length, since replicative senescence is more likely when telomeres fall below a critical limit. Telomere length and replicative capacity in sheep fibroblasts can be restored by transfecting the cells with the human telomerase gene, extending the lifespan by over 400 divisions[1]. Recently Cui and colleagues (32) have demonstrated that such telomase-immortalized cells are competent for nuclear transfer and can recapitulate early development. (See references (23–32).)

6.10   Gene targeting in humans

Zwaka and Thomson (33) have recently achieved gene targeting in human ES cells. They successfully disrupted both a selectable locus (*HPRT1*) and a non-selectable locus (*OCT4*) by electroporation in separate experiments, and recovered ES cells that appeared morphologically and physiologically normal. Both loci are expressed in normal human ES cells so a promoterless-type vector was suitable for the targeting experiment. It has yet to be established whether gene targeting can be achieved in non-expressed genes. (See reference (33).)

---

1 The telomerase-immortalized cells underwent 450 divisions before the experiment was halted, with no sign of impending senescence. The control, untreated cells senesced after 50–60 divisions.

6.11   Site-specific recombination

Although gene targeting by homologous recombination is an extremely powerful technique, it can be applied on a routine basis only in mice. As discussed above, techniques for gene targeting in other mammals are very labor intensive, due to the inefficiency of gene targeting in most somatic cells and the requirement for nuclear transfer. Another disadvantage of gene targeting by homologous recombination is that the targeted gene is modified in the germline, so all cells are similarly affected at all stages of development through to maturity. Site-specific recombination addresses both these difficulties by extending the possibility of gene targeting to virtually any animal that is amenable to transgenesis and allowing the components required for recombination to be supplied in a restricted manner, making it possible to generate transgenic organisms in which transgenes can be conditionally modified. In mice, the use of these methods in combination with gene targeting allows the production of conditional mutants (*conditional knockouts*), in which an endogenous gene is inactivated specifically in certain cell types or at a particular stage of development.

As discussed above, site-specific recombination differs from homologous recombination in its requirement for short, specific sequences and its restriction as a naturally occurring process to certain organisms. In terms of gene transfer, this means that target sites for site-specific recombination can be introduced into transgenes without major disruption to the construct, and that recombination will only occur in a heterologous cell if a source of recombinase is also supplied. The power of site-specific recombination as a tool for genome manipulation thus relies on the ability of the experimenter to supply the recombinase enzyme on a conditional basis.

A number of different site-specific recombination systems have been identified and several have been studied in detail. The most useful systems for gene transfer in animals are Cre recombinase from bacteriophage P1 and FLP recombinase (flippase) from the yeast *Saccharomyces cerevisiae*. These have been shown to function in many heterologous eukaryotic systems including mammalian cells and transgenic animals. Both the Cre and FLP enzyme recognize 34-bp sites (termed *loxP* and *FRP*, respectively) which contain minimally a pair of 13 bp inverted repeats surrounding an 8 bp central element. Cre recombinase has been used most extensively in mammalian systems because it has a temperature optimum of 37°C, while FLP functions optimally at 30°C and is preferred for *Drosophila*. (See references (34–38).)

6.12   Applications of site-specific recombination

The structure of the *loxP* site and the reaction catalyzed by Cre recombinase are shown in *Figure 6.7*. If two *loxP* sites are arranged as direct repeats, the recombinase will delete any intervening DNA leaving a single *loxP* site remaining in the genome. If the *loxP* sites are inverted with respect to each other, the intervening DNA segment is inverted. Although both reactions are reversible, excision reactions usually proceed to completion because the circular excised fragment is lost and degraded. However, as discussed later, this excised circle is very useful for genetic engineering in *Drosophila* (see Section 6.14).

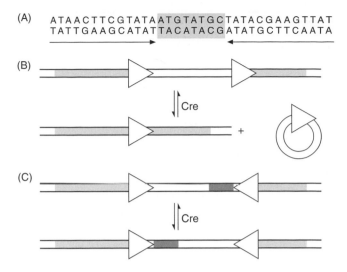

(A) ATAACTTCGTATAATGTATGCTATACGAAGTTAT
     TATTGAAGCATATTACATACGATATGCTTCAATA

Figure 6.7

(A) The sequence of the *loxP* site, recognized by Cre recombinase. (B) The reaction catalyzed by Cre recombinase if two *loxP* sites flank a particular DNA sequence as direct repeats. The DNA sequence is excised as a circle, which is rapidly degraded. (C) The reaction catalyzed by Cre recombinase if two *loxP* sites flank a particular DNA sequence as inverted repeats. The intervening DNA sequence is inverted.

One application of Cre-mediated deletion is the removal of unwanted sequences, such as selectable marker genes, after transformation. This provides a much simpler method for the production of mice carrying subtle mutations compared to the strategies discussed in Section 6.6, each of which requires two consecutive rounds of homologous recombination. In the Cre-dependent method, the second round of homologous recombination is unnecessary. As shown in *Figure 6.8*, gene targeting is used to replace the wild-type allele of a given endogenous gene with an allele containing a point mutation in a chosen exon, and simultaneously introduce markers, such as *neo* and *Tk* for positive and negative selection, into an adjacent intron. These markers are flanked by *loxP* sites. A second negative marker (e.g. the gene for diphtheria toxin) is included outside the homology region to select against random integration events. Cells that have lost the diphtheria toxin gene and also survive selection for *neo* are likely to have undergone authentic targeting events. Such cells are then transiently transfected with a plasmid expressing Cre recombinase, which catalyzes the excision of the remaining markers, leaving a clean point mutation and no evidence of genome modification except for a single *loxP* site in the intron, which should have no effect on the reconstituted gene once the markers have been removed. Negative selection against *Tk* using ganciclovir or FIAU identifies cells that have lost the markers.

Site-specific recombination can also be used to activate transgenes which have been introduced into cells as inactive constructs. In one method, termed *recombinase activated gene expression* (RAGE), a polyadenylation site

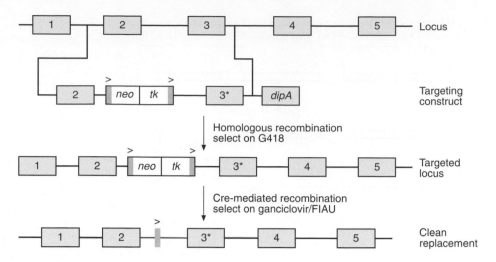

Figure 6.8

Gene targeting followed by marker excision, catalyzed by Cre recombinase. Initially, positive and negative markers (*neo* and *Tk*), flanked by *loxP* sites, are introduced into an intron by homologous recombination while a desired mutation is introduced into the adjacent exon. A second negative marker, in this case encoding diphtheria toxin, is included outside the homology region to eliminate random integration events. Following selection for *neo* on G418, Cre recombinase is used to excise both markers, leaving a single *loxP* site remaining in the intron. The excision event can be identified by selection for the absence of *Tk*, using ganciclovir or FIAU. Asterisk indicates mutation.

is placed between the transgene and its promoter (*Figure 6.9*). This prevents transgene expression by terminating any nascent transcripts before the transcription apparatus reaches the transgene. However, by placing *loxP* sites on either side of the polyadenylation site, Cre recombinase can be used to excise it and activate the transgene. The conditional provision of the recombinase allows the transgene to be activated in selected cells or at a specific developmental time point.

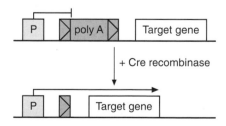

Figure 6.9

Overview of the recombinase activation of gene expression (RAGE) strategy. A polyadenylation signal is inserted between the promoter and target gene, blocking its expression. However, if this signal is flanked by *loxP* sites, Cre recombinase can be used to excise the block, bringing the promoter and gene into juxtaposition and thus activating gene expression.

Site-specific recombination can be used to introduce transgenes at specific genomic sites if a recombinase sequence such as *loxP* is already present. The use of an unmodified *loxP* site favors re-excision of the integrated transgene, even if Cre is supplied transiently. Therefore, Feng and colleagues (41) developed a system in which the *loxP* sites on the transgene are mutated in such a way that integration was possible but excision was not. Site-specific transgene integration into mammalian cells has also been achieved using FLP recombinase.

As well as rearrangements within individual genes, site-specific recombination between recombinase target sites that have been introduced at different genomic locations can be used to generate translocations, deletions and inversions that model naturally occurring chromosome disorders. Although this type of experiment was first achieved in *Drosophila* by Golic (42) using FRP recombinase, several analogous experiments using Cre recombinase in mice have also been reported. Desired rearrangement events can be selected if the recombination event reconstitutes a split selectable marker gene. For example, this has been achieved by providing complementary but non-functional fragments of *Hprt* at each targeted locus, and then driving recombination by transient expression of a Cre-expressing vector to restore *Hprt* function. (See references (15, 39–44).)

6.13   Conditional mutagenesis in mice

Conditional mutants display their mutant phenotype under defined (restrictive) conditions but not under other (permissive) conditions, and this is an established method used in microbiology to study gene products whose constitutive loss is lethal. As stated above, conventional gene targeting in mice results in constitutive mutations and in some cases this means that gene knockout experiments are uninformative, because the loss of function is lethal in early development.

To circumvent this problem, gene targeting and site-specific recombination can be combined to generate *conditional knockout mutants*. Targeting vectors are designed so that *loxP* sites are introduced into an endogenous gene, such that they flank an essential exon. Such a gene is said to be *floxed*. Cre recombinase is then supplied under the control of a cell type-specific, developmentally regulated or inducible promoter, causing the gene segment defined by the *loxP* sites to be deleted in cells or at the developmental stage specified by the experimenter (*Figure 6.10*). A widely used strategy is the use of two separate transgenic lines, one expressing Cre-recombinase in a regulated manner, and one containing the target gene which has been floxed. Crosses between these lines bring the *cre* gene and its target together in the same mouse, and as soon as Cre is activated the deletion event occurs in the appropriate cells. Cre-expressing transgenic mice are commercially available with Cre expressed under a wide range of different promoters, allowing gene targeting or transgene manipulation in any conceivable spatial or temporal pattern. The same Cre-transgenics can be used to conditionally activate or inhibit transgenes, or generate condition chromosome modifications.

The first report of this conditional knockout approach was published in 1994 by Gu and colleagues (45) and involved a Cre-transgenic line expressing

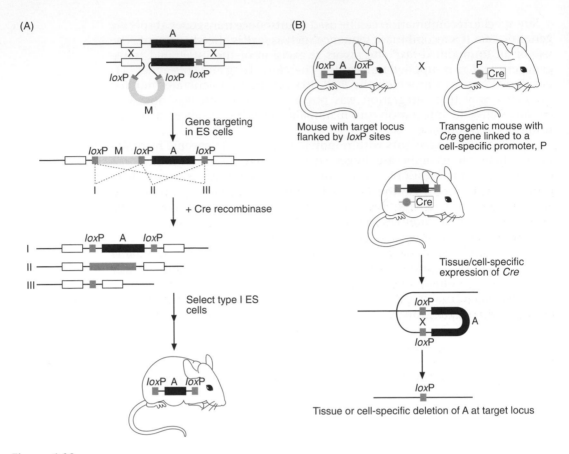

Figure 6.10

Gene targeting with the Cre-*loxP* recombination system can be used to inactivate a gene in a desired cell type. (A) Illustration of a standard homologous recombination method using mouse ES cells, in which three *loxP* sites are introduced along with a marker M at a target locus A (typically a small gene or an internal exon which if deleted would cause a frameshift mutation). Subsequent transient expression of a Cre recombinase gene results in recombination between the introduced *loxP* sites to give different products. Type 1 recombinants are used to generate mice in which the target locus is flanked by *loxP* sites. Such mice can be mated with previously constructed transgenic mice (B) which carry an integrated construct consisting of the Cre recombinase gene linked to a tissue-specific promoter. Offspring which contain both the *loxP*-flanked target locus plus the *cre* gene will express the *cre* gene in the desired cell type, and the resulting recombination between the *loxP* sites in these cells results in tissue-specific inactivation of the target locus. Taken from Strachan *et al*, *Human Molecular Genetics*, Copyright (2004), Garland Science.

the recombinase under the control of the T-cell-specific *lck* promoter, and a responder line in which an essential exon of the DNA polymerase β gene was floxed. Kuhn *et al.* (46) were the first to express Cre under an inducible promoter in mice – this was the metallothionein promoter described in Chapter 7, which is responsive to heavy metals and interferons. Cre has also been expressed as an estrogen–receptor fusion, making it inducible by the estrogen analog tamoxifen (page 187). A database of Cre transgenic mice and a database of floxed genes are maintained by the Nagy Lab and can be accessed at the following URL: http://www.mshri.on.ca/nagy/cre.htm. Another useful resource is T-BASE, a database of transgenic and knockout animals (flies, mice, rats, pigs) maintained by the Jackson Laboratory (http://tbase.jax.org). (See references (15,39,40,45).)

6.14 Gene targeting in *Drosophila*

As discussed in Chapter 5, the transformation of flies is normally achieved by the injection of P-elements, borne on plasmid vectors, into the pole plasm region of early embryos. The direct injection of DNA does not result in gene targeting even if the construct is homologous to an endogenous gene, and this is because the P-element functions specifically as an integrating element. Although rare targeting events have been recovered by selection in a *Drosophila* somatic cell line, flies have no cell system equivalent to ES cells for the production of transgenics.

To overcome this problem, gene targeting in *Drosophila* is accomplished by generating the donor molecule *in vivo*, which is achieved using a novel combination of endonuclease digestion and site-specific recombination. Three P-elements are required, initially introduced into three different fly lines and then stacked by two generations of crossing. The first P-element contains the gene encoding the yeast site-specific recombinase FLP, which excises DNA sequences between direct repeats of the *FRT* site as a circular product, much as shown for Cre-mediated excision in *Figure 6.7B*. The second P-element contains the gene encoding the rare-cutting endonuclease I-*Sce*1, which has a 22-bp recognition site not found anywhere in the wild-type *Drosophila* genome. Both genes are under the control of a heat-shock promoter. The final element contains the sequence of the donor molecule, which is homologous to the intended target site elsewhere in the genome, together with an adjacent marker, the *white* gene, which produces flies with the wild-type red eye color in a background of flies with white eyes[2]. The donor sequence is interrupted by the recognition site for I-*Sce*1, and the donor sequence and marker are flanked by *FRT* sites (*Figure 6.11A*). When all three elements are present in the same fly line, the flies are heat-shocked. The FLP enzyme then excises the donor molecule as a circular element containing a marker gene, homology region and a single FRT site. Simultaneously, the I-Sce1 enzyme cleaves the donor molecule within the homology region, producing a linear construct analogous to the insertion-type element shown in *Figure 6.1*. The donor sequence can then align with its target and undergo homologous recombination, resulting in the targeted disruption of the endogenous gene (*Figure 6.11B*).

To recover targeting events, the heat-shocked flies are mated to wild types. Most progeny of this cross lack the marker gene because the FLP-excised donor molecule is often lost. However, where targeting has been successful, the marker will be present in the offspring and will display the corresponding phenotype. Each stage of the targeting process can be

---

2 Remember, *Drosophila* genes are named for their mutant phenotype. The *white* gene encodes an enzyme in the pigment biosynthesis pathway that makes eyes red. In a *white* mutant background (where the endogenous *white* gene is non-functional) the use of a functional *white* gene as a marker produces flies with red eyes.

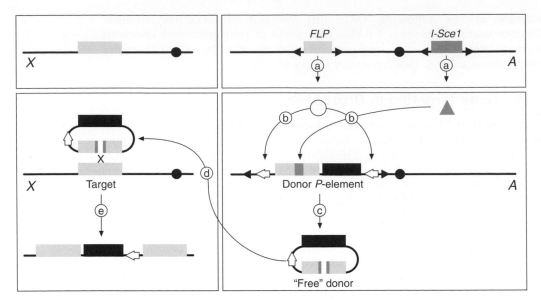

Figure 6.11

Gene targeting in *Drosophila*. In this example, the target locus is on the X-chromosome (X) and is identified by a grey box (top left panel). Flies with three P-elements are made by crossing independently transformed lines. In this example, the elements containing the *flp* and *I-sce1* transgenes are present on the same autosome (A) (top right panel) while the element containing the donor sequence is present on another autosome (bottom right panel). Expression of *flp* and *I-sce1* transgenes (a) results in the production of the corresponding enzymes, which act on the donor construct (b) producing a linearized free donor molecule (c). In the bottom left panel, this is shown synapsing with the target locus (d) and recombining to generate a duplicated target gene interrupted by the marker and one *FRP* site. White circle is FLP enzyme, white arrows are *FRP* sites, grey triangle is I-Sce1, dark grey bar is I-*Sce*1 target site, black box is marker gene, black arrows represent boundaries of P-elements. Reprinted from Biochemical and Biophysical Research Communications, Vol 297, No. 1, Rong, Y. "*Gene Targeting by . . .*", Pages 1–5, Copyright (2002), with permissions from Elsevier.

confirmed in different ways. Successful excision of the donor sequence can be confirmed by FLP expression, since only the intact donor element has the marker gene flanked by two FRT sites. However, the marker will survive FLP-mediated excision through random integration as well as homologous recombination. Therefore, genuine targeting events must be confirmed by genetic linkage analysis and molecular analysis, such as PCR or Southern blot hybridization. As with gene targeting in mammals, the efficiency of targeting in *Drosophila* is both locus dependent and related to the length of the homology region. Unlike the situation in mammals, however, there are only ever a maximum of two donor molecules in the cell (if the cell is in G2) rather than the hundreds or thousands introduced into mammalian cells by transfection. This suggests that in vivo donors are far more efficient substrates for homologous recombination than exogenously supplied DNA.

As is the case for gene targeting in mice, the *Drosophila* targeting strategy leads to a duplication of the homology region. However, because *Drosophila* genes can be small compared to those in mammals, providing

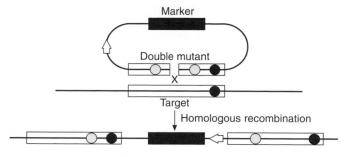

Figure 6.12

Scheme for generating targeted mutants in *Drosophila*. The targeting construct contains a point mutation either side of the I-*Sce*1 recognition site, so that insertion by homologous recombination generates two copies of the target gene, each containing a different mutation. Spontaneous intrachromosomal recombination to resolve such duplications has not been observed in *Drosophila*, although the presence of two mutations would prevent reversion to wild type should such an event occur. Reprinted from Biochemical and Biophysical Research Communications, Vol 297, No. 1, Rong, Y. "*Gene Targeting by . . .*", Pages 1–5, Copyright (2002), with permissions from Elsevier.

enough homology to facilitate efficient recombination may in some cases require the entire gene to be present. Where it is possible to use a small internal gene fragment as the homology region, the result of gene targeting is similar to that shown in *Figure 6.1*, where the result is two deficient copies of the gene, one truncated at the 5′ end and one truncated at the 3′ end. Various schemes have been devised to ensure mutation of the target locus where the homology region encompasses the entire gene. The one shown in *Figure 6.12* involves the creation of two point mutations, one either side of the I-*Sce*1 site. This produces a tandem duplication of the entire target gene separated by the marker, each copy harboring a different mutation. (See references (47,48).)

References

1. Smithies O, Gregg RG, Boggs SS, Koralewski MA, Kucherlapati R (1985) Insertion of DNA sequences into the human b-globin locus by homologous recombination. *Nature* **317**: 230–234.
2. Thomas KR, Capecchi MR (1987) Site-directed mutagenesis by gene targeting in mouse embryo-derived stem cells. *Cell* **51**: 503–512.
3. Soriano P (1995) Gene targeting in ES cells. *Annu Rev Neurosci* **18**: 1–18.
4. Ramirezsolis R, Davis AC, Bradley A (1993) Gene targeting in embryonic stem-cells. *Method Enzymol* **225**: 855–878.
5. Winding P, Berchtold MW (2001) The chicken B cell line DT40: a novel tool for gene disruption experiments. *J Immunol Methods* **249**: 1–16.
6. Hasty P, Bradley A (1993) Gene targeting vectors for mammalian cells. In: Joyner A (ed) *Gene Targeting – A Practical Approach*, pp 1–32. IRL Press, Oxford, UK.
7. Deng C, Capecchi MR (1992) Re-examination of gene targeting frequency as a function of the extent of homology between the targeting vector and the target locus. *Mol Cell Biol* **12**: 3365–3371.

8. Te Riele H, Maandag ER, Berns A (1992) High efficiency gene targeting in embryonic stem cells through homologous recombination with isogenic DNA constructs. *Proc Natl Acad Sci USA* **89**: 5128–5132.

9. Yang Y, Seed B (2003) Site-specific gene targeting in mouse embryonic stem cells with intact bacterial artificial chromosomes. *Nature Biotechnol* **21**: 447–451.

10. Mansour SL, Thomas KR, Capecchi MR (1988) Disruption of the proto-oncogene int-2 in mouse embryo-derived stem cells: a general strategy for targeting mutations to non-selectable genes. *Nature* **336**: 348–352.

11. Schwartzberg PL, Goff SP, Robertson EJ (1989) Germ line transmission of a c-abl mutation produced by targeted gene disruption in ES cells. *Science* **246**: 799–803.

12. Hasty P, Ramirez-Solis R, Krumlauf R, Bradley A (1991) Introduction of a subtle mutation into the Hox 2.6 locus in embryonic stem cells. *Nature* **350**: 243–246.

13. Moore RC, Redhead NJ, Selfridge J, Hope J, Manson JC, Melton DW (1995) Double replacement gene targeting for the production of a series of mouse strains with different prion protein gene alterations. *Biotechnology* **13**: 999–1104.

14. Capecchi MR (1989) The new mouse genetics – altering the genome by gene targeting. *Trends Genet* **5**: 70–76.

15. Muller U (1999) Ten years of gene targeting: targeted mouse mutants, from vector design to phenotype analysis. *Mech Develop* **82**: 3–21.

16. Papaioannou V, Johnson R (1993) Production of chimeras and genetically defined offspring from targeted ES cells. In: Joyner A (ed) *Gene Targeting – A Practical Approach*, pp 107–146. IRL Press, Oxford, UK.

17. Zheng B, Vogel H, Donehower LA, Bradley A (2002) Visual genotyping of a coat color tagged p53 mutant mouse line. *Cancer Biol Ther* **1**: 433–435.

18. Porteous MH, Cathomen T, Weitzman MD, Baltimore D (2003) Efficient gene targeting mediated by adeno associated virus and double-strand breaks. *Mol Cell Biol* **23**: 3558–3565.

19. Miller DG, Petek LM, Russell DW (2003) Human gene targeting by adeno-associated virus vectors is enhanced by DNA double-strand breaks. *Mol Cell Biol* **23**: 3550–3557.

20. Yanez RJ, Porter AC (2002) Differential effects of Rad52p overexpression on gene targeting and extrachromosomal recombination in a human cell line. *Nucl Acids Res* **30**: 740–748.

21. Dominguez-Bendala J, Priddle H, Clarke A, McWhir J (2003) Elevated expression of exogenous Rad51 leads to identical increases in gene-targeting frequency in murine embryonic stem (ES) cells with both functional and dysfunctional p53 genes. *Exp Cell Res* **286**: 298–307.

22. Semionov A, Cournoyer D, Chow TY (2003) 1,5-isoquinolinediol increases the frequency of gene targeting by homologous recombination in mouse fibroblasts. *Biochem Cell Biol* **81**: 17–24.

23. Denning C, Priddle H (2003) New frontiers in gene targeting and cloning: success, application and challenges in domestic animals and human embryonic stem cells. *Reproduction* **126**: 1–11.

24. Edwards JL, Schrick FN, McCracken MD, van Amstel SR, Hopkins FM, Welborn MG, Davies CJ (2003) Cloning adult farm animals: a review of the possibilities and problems associated with somatic cell nuclear transfer. *Am J Reprod Immunol* **50**: 113–123.

25. McCreath KJ, Howcroft J, Campbell KH, Colman A, Schnieke AE, Kind AJ (2000) Production of gene-targeted sheep by nuclear transfer from cultured somatic cells. *Nature* **405**: 1066–1069.

26. Denning C, Burl S, Ainslie A, *et al.* (2001) Deletion of the alpha(1,3)galactosyl transferase (GGTA1) gene and the prion protein (PrP) gene in sheep. *Nature Biotechnology* **19**: 559–562.

27. Dai Y, Vaught TD, Boone J *et al.* (2002) Targeted disruption of the alpha1,3-galactosyltransferase gene in cloned pigs. *Nature Biotechnology* **20**: 251–255.

28. Lai L, Kolber-Simonds D, Park KW, *et al.* (2002) Production of alpha-1,3-galactosyltransferase knockout pigs by nuclear transfer cloning. *Science* **295**: 1089–1092.

29. Phelps CJ, Koike C, Vaught TD, *et al.* (2003) Production of alpha 1,3-galactosyltransferase-deficient pigs. *Science* **299**: 411–414.

30. Rubin H (2002) The disparity between human cell senescence in vitro and life-long replication in vivo. *Nature Biotechnology* **20**: 675–681.

31. Cui W, Aslam S, Fletcher J, Wylie D, Clinton M, Clark AJ (2002) Stabilisation of telomere length and karyotypic stability are directly correlated with the level of hTERT gene expression in primary fibroblasts. *J Biol Chem* **277**: 38531–38539.

32. Cui W, Wylie D, Aslam S, Dinnyes A, King T, Wilmut I, Clark AJ (2003) Telomerase immortalised fibroblasts can be reprogrammed by nuclear transfer to undergo early development. *Biol Reprod* **69**: 15–21.

33. Zwaka TP, Thomson JA (2003) Homologous recombination in human embryonic stem cells. *Nat Biotechnol* **21**: 319–321.

34. Gossen M, Bujard H (2002) Studying gene function in eukaryotes by conditional gene inactivation. *Annu Rev Genet* **36**: 153–173.

35. Craig NL (1988) The mechanism of conservative site-specific recombination. *Annu Rev Genet* **22**: 77–105.

36. Sadowski PD (1993) Site-specific genetic recombination: hops, flips and flops. *FASEB J* **7**: 760–767.

37. Sauer B (1994) Site-specific recombination: developments and applications. *Curr Opin Biotechnol* **5**: 521–527.

38. Buchholz F, Ringrose L, Angrand PO, Rossi F, Stewart AF (1996) Different thermostabilities of FLP and Cre recombinases: implications for applied site-specific recombination. *Nucleic Acids Res* **24**: 4256–4262.

39. Metzger D, Feil R (1999) Engineering the mouse genome by site-specific recombination. *Curr Opin Biotechnol* **10**: 470–476.

40. Albanese C, Hulit J, Sakamaki T, *et al.* (2002) Recent advances in inducible expression in transgenic mice. *Semin Cell Dev Biol* **13**: 129–141.

41. Feng YQ, Seibler J, Alami R, Eissen A, Westerman KA, Lebouch P, Fiering S, Bouhassira EE (1999) Site-specific chromosomal integration in mammalian cells: Highly efficient CRE recombinase-mediated cassette exchange. *J Mol Biol* **292**: 779–785.

42. Golic K (1991) Site-specific recombination between homologous chromosomes in *Drosophila*. *Science* **252**: 958–961.

43. Ramirez-Solis R, Lui P, Bradley A (1995) Chromosome engineering in mice. *Nature* **378**: 720–724.

44. Li ZW, Stark G, Gotz J, Rulicke T, Gschwind M, Huber G, Muller U, Weissmann C (1996) Generation of mice with a 200-kb amyloid precursor protein gene deletion by Cre-recombinase-mediated site-specific recombination in embryonic stem cells. *Proc Natl Acad Sci USA* **93**: 6158–6162.

45. Gu H, Marth JD, Orban PC, Mossmann H, Rajewsky K (1994) Deletion of a DNA polymerase beta gene segment in T-cells using cell type-specific gene targeting. *Science* **265**: 103–106.

46. Kuhn R, Schwenk F, Auget M, Rajewsky K (1995) Inducible gene targeting in mice. *Science* **269**: 1427–1429.

47. Rong YS (2002) Gene targeting by homologous recombination: a powerful addition to the genetic arsenal for *Drosophila* geneticists. *Biochem Biophys Res Commun* **297**: 1–5.

48. Gloor GB (2001) Gene targeting in *Drosophila* validated. *Trends Genet* **17**: 549–551.

Protocols

CONTENTS

MATERIALS

Vasectomized male mice (see Protocol 5.1)

Superovulated female mice, strain C57BL/6 (see Protocol 5.3)

Pseudopregnant foster mothers (see Protocol 5.5)

Micromanipulation apparatus (see Materials, Chapter 5) with beveled injection pipette

Targeting construct

3 M sodium acetate

Ethanol (70%, absolute)

PBS

ES cells

Tissue culture flasks coated in gelatin

6-, 12- and 48-well plates, coated in gelatin

DMEM + 10% fetal bovine serum

Media additives recommended for the ES cell line being used (e.g. LIF)

Trypsin

Hemocytometer

Bio-Rad Gene Pulser and electroporation cuvettes

Geneticin (G418)

Ganciclovir (optional)

Injection medium (DMEM + 25 mM HEPES without $NaHCO_3$, pH 7.4)

Protocol 6.1: Electroporation of murine ES cells with G418 selection

1. Prepare high-quality DNA by alkaline lysis and cesium chloride density centrifugation or using a column purification system such as a QIAGEN maxi-prep.

2. Linearize 150 µg of vector DNA overnight at 37°C with the appropriate restriction enzyme in 0.5 ml final volume.

3. Precipitate the DNA sample with 0.1 volumes of 3 M sodium acetate and 2.5 volumes of absolute ethanol on ice for 5 min. Pellet the DNA at 13 000 g with a bench microfuge, wash twice with 70% ethanol and allow to dry.

4. Resuspend the pellet in 0.1 ml sterile PBS. Store at –20°C until needed.

5. Two days before transfection seed ES cells into 175 cm^2 flasks using the recommended culture medium (this can be obtained from the supplier of the cells). Split the cells when they become 80% confluent and return to the incubator.

6. When the cells reach 80% confluence, change the medium and return to the incubator for 2 hours. Then pour off the medium from three flasks and trypsinize the contents using 5 ml of the trypsin solution per flask. Add 10 ml of medium to each flask and combine the floating cells in a 50 ml sterile centrifuge tube. Pellet cells for 5 minutes at 1200 rpm and resuspend in 20 ml PBS. Meanwhile allow an aliquot of the linearized vector to defrost.

7. Determine the cell density by counting a small volume in a hemocytometer. Pellet as above and resuspend in a volume sufficient to yield a cell density of 1.5×10^8 ml^{-1}.

8. Transfer 0.7 ml of cells to the tube containing 0.1 ml of linearized vector DNA, quickly mix with a 1 ml plastic pipette and transfer immediately to a 0.4 cm Bio-Rad electroporation cuvette. Electroporate using the Gene Pulser instrument at 3 µF/800V (time constant = 0.04 msec).

9. Allow the cells to recover for 20 minutes at room temperature and then transfer them to 200 ml of fresh medium. Plate 10 ml (approximately 5×10^6 cells) onto each of 20 gelatinized 10 cm tissue-culture dishes.

10. The following day, remove the medium and replace with fresh supplemented with 125 µg ml^{-1} G418. If a *Tk* marker is present on the vector, ganciclovir can also be included at this point to eliminate random integration events. Change the medium daily for the first 4–5 days. Once most of the cells have been cleared from the plate, the medium may be replaced on alternate days. Eight days after electroporation, the colonies should be about 1 mm in diameter.

Protocol 6.2: Picking targeted colonies

1. Under a laminar flow hood, remove the medium on the ES cells and carefully examine the bottom of each dish, circling colonies with a marker pen. Gelatinize an appropriate number of 48-well plates.

2. Add 10 ml of room temperature PBS to each dish. Break up each colony with a sterile pipette tip and collect in 50 μl of PBS. Transfer the cells to individual wells in the 48-well plate.

3. Add 50 μl of a 2× trypsin solution to each well and incubate for 10 minutes at 37°C. Tap plates to disperse the cells into small clumps and add 1 ml of medium to each well. The following day, refresh with 0.5 ml of medium.

4. When the cells approach confluence, trypsinize with 100 μl of 2× trypsin for 2–4 minutes and then add 1 ml of medium. Remove 400 μl of each suspension to a labeled microfuge tube and use a QIAGEN or similar miniprep kit to extract the DNA and test for targeting by PCR. The remaining cells should be replaced in the incubator.

5. Correctly targeted cells can be expanded into larger plates for freezing. Maintain selection throughout the expansion process.

Protocol 6.3: Blastocyst injection

1. Thaw cells from appropriate positive clones about 5 days prior to injection. Seed into dishes and allow to become 80% confluent. Split cells into several dishes if injections will extend over several days.

2. Refresh the medium 2 hours before the injections, but do not include selection agents from now on. Just before transfer, trypsinize the cells for 5 minutes with prewarmed trypsin at 37°C. Place the trypsinized cells in a 15 ml tube, and add an equal volume of medium to inactivate the trypsin. Centrifuge the cells briefly and remove the supernatant. Resuspend the cells in 1 ml of injection medium + 10% fetal bovine serum and pipette gently to fully separate the cells.

3. Place the cells on ice for 1–2 hours to prevent clumping.

4. Flush blastocysts from pregnant C57BL/6 females and collect into DMEM + 10% fetal bovine serum. The procedure for operating on the mice is initially the same as described in Protocol 5.4, although the operation takes place on the fourth day of pregnancy (3.5 days post coitus). Once the uterine horns have been removed and separated from the cervix and ovary, transfer to a round-bottomed watchglass and use a Pasteur pipette to flush about 0.5 ml of medium through each horn from the cervical end. The embryos will be flushed through and will sink to the bottom of the glass. Transfer to fresh medium and allow to expand for 1–2 hours in a 37°C incubator with 6% CO_2. Transfer the blastocysts to the injection chamber precooled to 4°C. Transfer ES cells into the chamber, adjacent to the blastocysts.

5. The micromanipulation apparatus for blastocyst injection is similar to that used for pronuclear microinjection, except that the injection needle has a wider bore to permit the passage of cells without damage, and it is beveled. Take up 20–30 cells. Inject each blastocyst with sufficient cells to fill the blastocele (usually 15–20). Then place blastocysts into pseudopregnant recipient females (10–15 blastocysts per uterine horn) using the procedure for implanting injected eggs (Protocol 5.5). For every 10 blastocysts, 1–2 will produce male chimeras with germline mosaicism.

Control of transgene expression

7

7.1   Introduction

In many, probably most, gene transfer experiments, the aim is to express the transgene or transgenes and produce some kind of gene product. We have already encountered some examples where this is not the case, e.g. the gene-targeting strategies discussed in Chapter 6, and we shall consider some more in Chapter 9 when we describe how gene transfer is used for insertional mutagenesis. However, it is fair to say that in most cases, the experimenter needs to make some provision to control the expression of the introduced genes, which means it is necessary to provide those genes as part of an *expression construct* containing the information required for transcription, RNA processing, translation and in many cases post-translational trafficking of the product. Constitutive expression is suitable for many experiments in cultured cells, whereas in transgenic animals it is often more appropriate to restrict transgene expression to particular cell types, tissues or stages of development. In both cells and animals, more precise control of transgene expression can be achieved through the use of inducible expression systems, in which the transgene or its product is brought under the control of a small chemical inducer. In this chapter we discuss the principles of gene expression control as applied in transgenic systems and look at the different regulatory elements that are available for expression construct design.

7.2   Classes of expression vector

The conventional expression vector is a construct containing the transgene coding region and a series of regulatory elements that allow that gene to be expressed in the appropriate host cell to yield either a functional RNA sequence or a recombinant protein. However, there are several different forms of expression vector suited to different purposes. The simplest expression vectors, *transcription vectors*, allow transcription but not translation of cloned foreign DNA, and are designed for use *in vitro* (*Figure 7.1*). They are used to synthesize large quantities of recombinant RNA (up to 100 μg per reaction) and are often equipped with dual, opposing promoters allowing both message sense and antisense RNA to be synthesized. Significantly, such vectors utilize the very specific promoters from *E. coli* bacteriophages such as T3, T7 and SP6, whose RNA polymerase enzymes can be supplied *in vitro*. These promoters do not function in animal cells unless the appropriate RNA polymerase is expressed as the product of another transgene. Such simple transcription vectors also tend to lack polyadenylation and transcriptional termination sites, so defined RNA molecules can be

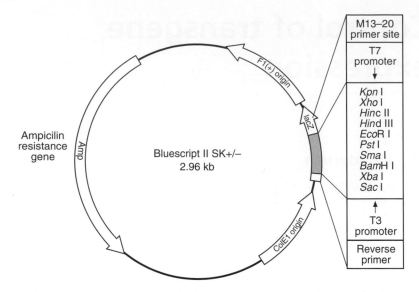

Figure 7.1

The general purpose cloning vector pBluescriptII SK +/–. Note the multiple cloning site (MCS) is flanked by opposing T7 and T3 promoters, which allow the run-off transcription of any cloned DNA sequence. Bluescript is a registered trademark of Stratagene. Image used with permission from Stratagene.

produced only by prior linearization, allowing the polymerase to run off the end of the template (*run-off transcription*). Other, more sophisticated transcription vectors are designed for use in cells or animals. Examples of these include vectors for the expression of ribozymes, antisense RNA, hairpin RNA and siRNAs (see Chapter 9). Such vectors contain promoters that function in animal cells as well as proper transcriptional termination signals.

Protein expression vectors allow both the transcription and translation of cloned DNA and may include additional sequences, which determine the destination of the protein after synthesis, thereby also controlling the nature of any compartment-specific post-translational modifications such as glycosylation (*Figure 7.2*). Where the objective of the experiment is simply to produce as much recombinant protein as possible, the vectors are designed to maximize both transcription and translation – these are often called *overexpression vectors*. Where maximized expression is undesirable or toxic to the host cell or animal, inducible regulatory elements may be used or (in animals) regulatory elements that restrict transgene expression to particular cell types or stages of development. Proteins may be expressed as native polypeptides or fusion proteins, the latter often to increase protein stability, facilitate protein purification or provide reporter functions for analysis (Chapter 1). A common fusion strategy used in cell lines is to express recombinant proteins with an N-terminal signal peptide. This allows the protein to be secreted from the host cell and collected from the growth medium without lysing the cells. (See references (1,2).)

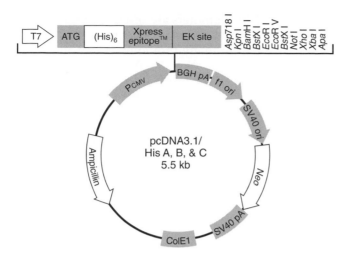

Figure 7.2

The transient expression vector pcDNA3.1/HisAB&C, marketed by Invitrogen Inc., which contains a typical mammalian expression cassette. This comprises the human cytomegalovirus immediate early promoter, bovine growth hormone polyadenylation site/transcriptional terminator site, and an ATG codon within an authentic Kozak consensus for efficient translation. The cassette also contains two epitope tags (His_6 and the proprietary Xpress epitope) separated from the multiple cloning site by an enterokinase (EK) protease target site allowing the epitopes to be removed from the recombinant protein after extraction and purification. The T7 promoter is also present allowing in vitro run-off transcription. Copyright (2004), Invitrogen Corporation. Used with permissions.

7.3 Components of protein expression constructs

Expression vectors are modular in nature and often contain a whole series of different elements with different functions. Some of these functions are required for cloning in bacteria. Indeed, most plasmid vectors for animal cells also have to replicate in bacteria and must therefore possess a bacterial origin of replication, plasmid maintenance and partition sequences and a bacterial selectable marker. Vectors for the stable transformation of animal cells may also contain a selectable marker for use in the animal cell host, while vectors for transient transfection are equipped with reporter genes to allow transfection to be confirmed and, if necessary, quantified. These sequences are discussed in more detail in Chapter 1. Another common feature of plasmid vectors for animal cells is a viral origin of replication. As described in Chapter 2, these allow the vectors to be maintained as episomal replicons. While important for vector preparation and introduction into the host, none of the above modules are directly involved in transgene expression.

The sequences that control transgene expression can be divided into several classes. Perhaps the most important are the promoter and enhancer sequences, since these define the expression profile and efficiency at the level of transcriptional initiation. The promoter is the site where RNA polymerase binds to the transcription unit (RNA polymerase II for most transgenes, but RNA pol III promoters are often used for the expression of short RNA molecules). As well as basal elements such as a TATA box and initiator, which are present in many promoters, there are a number of *cis*-acting elements upstream of the basal promoter to control cell type-specificity and induction in response to external signals. Enhancers are *cis*-acting elements that work with promoters to increase their transcriptional activity; they may also impart cell type- and inducible specificity upon gene expression. Promoter and enhancer systems are often derived from mammalian viruses because some viral regulatory elements are highly active and are functional in a wide range of cell types and species. Cell-specific and inducible promoter systems may be derived directly from endogenous genes, but inducible systems are often heterologous or entirely artificial to suppress background activity (see Section 7.4).

In addition to sequences controlling transcriptional initiation, further elements are required for the control of RNA processing. Such sequences generally include a transcriptional termination/polyadenylation site and, in many cases, an intron. Termination and polyadenylation sites are essential for stable RNA production. Introns are not essential in cell culture but they have been shown to increase the level of transgene expression for many unrelated genes. Introns do appear to be important for effective transgene expression in transgenic mice.

Sequences may also be included for efficient translation and protein targeting. The efficiency of protein synthesis is determined by many factors, including the stability of the mRNA, the structure of the non-coding regions and translational start site, and the codon usage within the transgene. The 5′ untranslated region (UTR) is generally kept as short as possible, since secondary structure in this part of the transcript can inhibit translational initiation. If a 3′ UTR is included in the vector, then AU-rich sequences should be removed because these often make the mRNA unstable. The Kozak sequence, which surrounds the initiation codon and helps to assemble the ribosome, should be matched as closely as possible to the ideal Kozak consensus. A further sequence included in some vectors is a picornaviral internal ribosome entry site, allowing the production and translation of polycistronic mRNAs. Sequences may also be included that target recombinant proteins to particular cellular compartments. Nuclear localization sequences are used to target proteins to the nucleus. Signal peptides target proteins to the secretory pathway, which may be essential for correct post-translational modification. KDEL retention signals (KDEL specifies a tetrapeptide sequence in the single letter amino acid code) may be used to favor intracellular protein accumulation. Importantly, none of the post-transcriptional regulatory elements used in standard expression vectors have any qualitative impact on transgene expression – only the promoter/enhancer system can direct transgene expression to particular cells and developmental stages or activate the transgene in response to external signals. The post-transcriptional elements only affect transgene expression quantitatively, by maximizing the efficiency of each stage of gene expression[1].

1 This is not to say that such elements do not exist, because many instances of cell-specific splicing and inducible regulation of mRNA stability and protein synthesis have been described. However, such elements are generally not included in standard expression vectors.

Fusion peptides/proteins may also be included in the expression vector to increase protein stability, facilitate protein purification or provide reporter functions. Fusion proteins such as glutathione-S-transferase and staphylococcal protein A are used as purification tags, facilitating affinity purification with glutathione and immunoglobulin G, respectively. Smaller purification tags include the epitopes c-Myc and FLAG, which are recognized by specific antibodies, and His_6 (six consecutive histidine residues), which can be purified by immobilized metal affinity chromatography. Fluorescent proteins and enzymes such as luciferase and β-galactosidase may be used as fusions to report protein abundance or localization, and β-galactosidase can also be exploited as a purification tag. The tag is often separated from the recombinant protein by a protease cleavage site, such as enterokinase or factor Xa. (See references (1,2).)

7.4 Transcriptional initiation: promoters and enhancers

Transcriptional initiation is the rate-limiting step in the expression of most eukaryotic genes. Control is mediated by *cis*-acting DNA elements that act as binding sites for *trans*-acting regulators termed *transcription factors*. Transcription factors can act either positively or negatively and in most cases interact directly with the basal transcriptional apparatus, thus either stimulating or inhibiting the recruitment and activity of RNA polymerase and the other basal components. Transcription factors can also influence chromatin structure or introduce kinks and bends into DNA, thereby controlling the structure of the template on which the initiation complex assembles. Some transcription factors act solely by influencing the activities of other transcription factors and do not interact with DNA or the basal transcription apparatus at all.

The complex expression patterns of animal genes reflect the combination of *cis*-acting elements present in the regulatory sequences and the availability of different transcription factors. All genes possess a *promoter*, located immediately 5′ to the transcriptional start site and many are also controlled by one or more *enhancers*, which can be located some distance away. The basal promoter is where the initiation complex assembles and recruits RNA polymerase. Basal promoters usually comprise an *initiator element* surrounding the transcriptional start site, and/or a motif termed the *TATA box* located about 25 bp upstream of the start site. Promoters may contain one or both of these elements, and some contain neither. Further motifs are located 5′ to these basal sequences and are known as *upstream promoter elements*. These may include sequences such as the CAAT box which bind to transcription factors found in all cells, as well as specialized regulatory motifs and response elements that respond to transcription factors that are either synthesized or activated in restricted spatiotemporal patterns. Similarly, enhancers are made up of different motifs that may act generally or may confer spatial, temporal or inducible specificity upon the gene. Because the basal promoter controls transcriptional initiation, it is always found on the same strand relative to the message sense of the gene and always in the same orientation to insure that transcription proceeds through the gene in one direction. Conversely, enhancers act over a distance by the looping out of intervening DNA and can be found on either strand and in either orientation relative to the gene, and may be located upstream, downstream or even within the gene. Enhancers may stimulate transcription up to 1000-fold from a given promoter, but have no intrinsic promoter activity themselves. (See references (3,4).)

Constitutive viral promoters

In the first gene transfer experiments, mammalian transgenes were expressed under the control of their own transcriptional control sequences, which were poorly defined and generally unpredictable in their activities. Therefore, researchers looked for regulatory elements that were strong and more reliable in a range of cell types, and turned to animal viruses with their simple genomes and wide host ranges. The cloning of viral regulatory elements provided the breakthrough in animal gene transfer, since transgenes could be expressed in many different host cells using a relatively small repertoire of promoters and enhancers. A small number of specific viral promoters have been relied upon to drive transgene expression in particular viral transduction vectors, including the adenoviral E1 promoter, the vaccinia virus p7.5 promoter and the baculovirus polyhedrin promoter (chapter 4). These have been chosen for a variety of reasons, including their strong activity and the fact that they control non-essential viral genes, which can therefore be replaced with a suitable transgene. For plasmid vectors, three viral promoter/enhancer systems have been extensively developed because of their impressive activity range: the SV40 early promoter and enhancer, the Rous sarcoma virus long terminal repeat promoter and enhancer and the human cytomegalovirus immediate early promoter. These elements actually function in a broader range of cells than defined by the host range of the viruses themselves, because although the appropriate receptor may be absent from the surface of the cell, the transcription factors are often present and can act upon viral sequences introduced by transfection. For example, SV40 can only infect simian cells and the SV40 origin of replication only allows episomal propagation in simian cells, but the promoter/enhancer system can function in mammalian cells of many species. However, none of these viral promoters is absolutely universal in its activity so there does appear to be some limitation to their usefulness. For example, SV40 sequences do not work well in HEK 293 cells, in which the adenoviral promoters are highly active. (See references (5–8).)

Endogenous spatially and temporally regulated promoters

While viral promoters have evolved to function in many cell types, the promoters of animal genes can be either constitutive or regulated. Constitutive promoters are usually found controlling the expression of so-called housekeeping genes, which are required for core metabolic and physiological functions. Other genes, sometimes called luxury genes, are regulated either in time or space or both. Promoters are available which restrict gene expression to particular cell types, to particular developmental stages, to stages of the cell cycle, to particular environmental conditions, and promoters are available which are active only in the presence of specific exogenous agents, drugs, metabolites, hormones, cytokines or growth factors. All these are useful under different experimental conditions and an entire book would be required to discuss them all in detail. The discussion below is restricted to those cell-specific promoters and (in the following sections) inducible promoters, which have been the most widely used in transgenic systems.

There is generally little need for cell-specific promoters in cultured cells, although these can be useful if the cell type is capable of differentiation ex vivo. For example, neuron-specific promoters have been used to drive gene expression in ES cells, since these can be induced to differentiate into neurons under appropriate culture conditions. In transgenic animals, cell-specific promoters are much more valuable and countless experiments have been described in which different cell-specific promoters have been used for specific experimental purposes. Arguably the most widely used cell-specific promoters, at least in a commercial setting, are those used to direct transgene expression to particular production organs, such as the mammary glad, urinary epithelium or reproductive organs. A number of such elements have been developed and protected by patents including the promoters for whey acidic protein, β-lactoglobulin, α_{S1} casein, β-casein, J-casein and α-lactalbumin. These promoters facilitate the production of recombinant proteins in milk at a level of several grams per liter. The mouse uroplakin II and uromodulin promoters have been used to direct transgene expression to the bladder epithelium in an alternative production system in which the protein is expressed in urine. Semen-specific promoters have also been used in pigs since pig seminal fluid is highly proteinacious and proteins expressed therein generally do not interfere with endogenous physiological processes. Another promoter which may become commercially valuable is the silkworm fibroin L-chain promoter, which can be used to produce large amounts of recombinant proteins in silkworm cocoons. (See references (3,4,9).)

Endogenous inducible promoters

High-level constitutive protein expression is suitable for many gene transfer experiments and applications, but where more subtle control is required, or where the recombinant product is toxic, it is useful to bring expression under exogenous control. The first such experiments utilized endogenous inducible promoters, three of which have been widely developed: the *Drosophila* heat shock promoter (responsive to elevated temperature), the mouse metallothionein promoter (responsive to the presence of heavy metals such as cadmium and zinc) and interferon-β promoter (responsive to interferon, viral infection and dsRNA). A further inducible promoter was discovered in the murine mammary tumor virus (MMTV) long terminal repeat, and was found to be responsive to dexamethasone, a synthetic glucocorticoid analog. A transgene under the control of one of these elements could be induced by changing the physical conditions under which the cells or animals were maintained (heat shock), or by adding the appropriate chemical or biological inducer to the culture medium or food/water. Unfortunately, all these systems suffered from one or more of the following five problems, which have therefore become checkpoints for the development of novel systems:

- Background activity, i.e. transgene activation in the absence of induction (metallothionein promoter especially)
- Poor induction ratio, i.e. transgene expression is not elevated significantly by induction (metallothionein promoter and MMTV promoter both have an induction ratio less than 10)
- Toxicity of the inducing agent (e.g. heat shock causes extensive cell death)
- Differential response to the inducing agent by different cells in a transgenic animal, reflecting in many cases differential penetration and recovery from the stimulus, and for chemical inducers different rates of absorption, distribution and metabolism (dexamethasone is a prime example)
- The activation of endogenous genes regulated by the same promoter (e.g. glucocorticoid-responsive genes are activated by dexamethasone).

(See references (10–14).)

Heterologous, recombinant and artificial inducible systems

Many of the problems attributed to the use of endogenous inducible systems can be addressed using systems that are heterologous or artificial, since the risk of activating non-target genes is minimal (reducing toxic effects) and the other properties – leakage, induction ratio, differential response in animals – can be modified by engineering the system and choosing more suitable inducing agents. In the search for an ideal inducible system, the following properties have been identified as important factors for success:

- Transgene induction is specific to an exogenously supplied, non-toxic agent
- Transgene induction does not interfere with any endogenous pathways because no endogenous genes are affected
- There is no background activity – in the non-induced state, the regulated transgene is silent
- There are no undesirable cell type-specific effects of induction due to differential uptake, distribution and metabolism of the inducer, but transgene expression can be induced in specific cell-types or at specific developmental stages by design if necessary
- The inducing agent penetrates cells quickly for rapid induction, and decays quickly for prompt return to the basal state.

One disadvantage is that since all the components of the system are foreign, not only the target transgene but also genes encoding all the appropriate regulatory components have to be introduced into the cell as additional transgenes. In transgenic animals, this situation is somewhat similar to Cre-*loxP* conditional mutagenesis as described in Chapter 6, and the problem can be addressed in the same way. That is, specific regulator strains can be developed which express the transcription factors of the system strongly and can be crossed to transgenic mice carrying the appropriate regulated transgene.

All the current inducible systems are based on a switch principle. Transgene expression is controlled by a response element which is not found elsewhere in the cell and is not influenced by any endogenous transcription factor. It is activated (or repressed) solely by a heterologous or artificial *trans*-acting regulator, whose activity is modulated by a small, exogenously supplied molecule (the inducer). Five major systems have been developed: the *Escherichia coli lac* operon, the *E. coli tet* operon, the *Drosophila* molting hormone ecdysone, the mammalian antiprogestin system and systems based on dimerizing drug analogs. Not all inducible switches function at the transcriptional level, and switches at the protein level are discussed later. (See reference (15).)

The *lac* and *tet* systems

The *E. coli lac* system is based on the *lacI* gene, which encodes a transcriptional repressor protein. The *lac* operon in *E. coli* encodes three enzymes involved in lactose metabolism. In the absence of lactose, the LacI repressor binds to operator sites around the transcriptional start site and prevents RNA polymerase initiating transcription. The presence of lactose in the medium results in the synthesis of a metabolic by-product, allolactose. This binds to the repressor, causing a conformational change that prevents the repressor from binding specifically to the operator sequences – instead the repressor now binds to any DNA sequence and is rapidly titrated out of the system. The same thing happens in the presence of the synthetic lactose analog IPTG (isopropyl-1-thio-β-D-galactopyranoside). Transcription of the genes for lactose catabolism then commences.

The strategy for Lac-based transgene regulation in animal cells is to insert one or more *lac* operator sequences downstream of the transgene transcriptional start site, and introduce a second expression construct constitutively expressing the *lacI* gene (*Figure 7.3*). Under basal conditions, the transgene is repressed, but if IPTG is added to the medium, the repressor is released from the operator site allowing transgene expression. This system achieves an induction ratio of about 3–500, which is up to 100-fold greater than MMTV and metallothionein systems. However, the number and position of *lac* operator sites can have a significant effect on the regulation of the target gene.

The original *E. coli tet* system was based on a similar principle to the *lac* system (*Figure 7.4*). The *tetR* gene encodes a repressor that binds to operator sequences in the *tet* operon when tetracycline is absent. Here, tetracycline itself can be the inducer although the synthetic analog doxycycline is used in most experiments. Like the impact of IPTG on the Lac repressor, doxycycline changes the conformation of the Tet repressor causing it to release the operator sequence and bind to any random DNA. The main difference between the *lac* and *tet* systems in mammalian cells is that the number and position of *tet* operator sequences is less important for inductive control than the equivalent *lac* operators. Despite the improvement over existing systems, these original *lac* and *tet* control circuits were often toxic in cells because high levels of the repressor proteins were required. Furthermore, IPTG and doxycycline are not ideal inducers since both tend to accumulate and their effects are slow to decay.

These problems were initially addressed by converting TetR and LacI from repressors into activators, which was achieved by fusing them to known

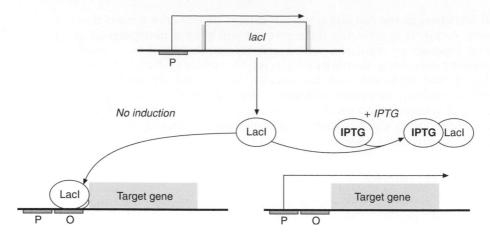

Figure 7.3

The *lac* repressor-based system for inducible transgene control. The transgene construct contains *lac* operator(s), which bind LacI repressor and prevent transcription. Induction is achieved with IPTG, which binds to and inactivates the repressor, causing it to release the DNA.

mammalian transactivation domains such as the herpes simplex virus VP16 gene product. In the case of the *tet* system, the fusion product (termed tTA, for *tet*-transactivator) confers constitutive transgene activity in the absence of doxycycline and is repressible by this antibiotic (hence the alternative name Tet-Off; *Figure 7.5*). Such systems have a reduced basal transcription level because high levels of the repressor protein are unnecessary, resulting in a much improved induction ratio of up to 10^5. One disadvantage,

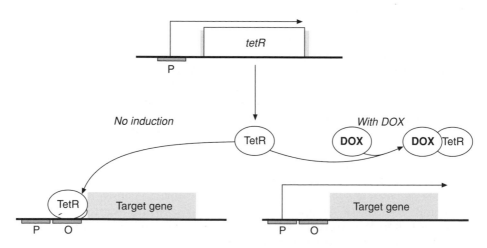

Figure 7.4

The *tet* repressor-based system for inducible transgene control. The transgene construct contains *tet* operator(s) which bind TetR repressor and prevent transcription. Induction is achieved with tetracycline, or its analog doxycycline (DOX), which binds to and inactivates the repressor, causing it to release the DNA.

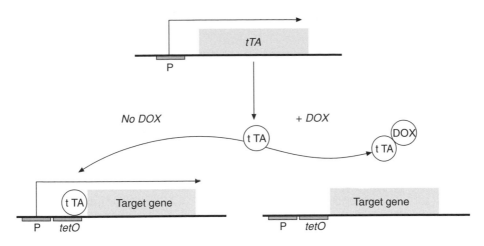

Figure 7.5

The tTA (tet-activator) system, in which the *E. coli* Tet repressor is fused to a transcriptional activation domain, e.g. *VP16*. The transgene contains *tet* operators that now function as enhancers. Introducing tetracycline or doxycycline into the system inactivates tTA causing transgene repression.

however, is the constant requirement for doxycycline in the culture medium or food/water to maintain the circuit in its basal state, which can be toxic to cells and animals because doxycycline accumulates. This also results in base level fluctuation in animals because of differential uptake and distribution of the antibiotic. Furthermore, since induction requires the absence of doxycycline, it is dependent on the rate at which this antibiotic can be eliminated, which also varies in different cells. The tTA protein is also toxic at high levels.

Further advances in *tet*-regulation technology have resulted in the creation of a mutated form of the tTA activator whose DNA-binding activity is improved rather than abolished by doxycycline, hence the alternative name Tet-On (*Figure 7.6*). This protein (termed rtTA, reverse *tet*-transactivator) is an activator in the presence of doxycycline, removing the requirement for long-term exposure to the inducer. Unfortunately, the mutations that reverse the effect of doxycycline also affect its binding activity, so up to ten times more antibiotic is required to activate rtTA than inhibit tTA. More recently, however, further mutated versions of tTA have been developed which do not affect doxycycline binding. The improved version is known as rtTA mark 2 (rtTA-M2). Another limitation of the original rtTA system is that the regulator displays a significant residual binding to the *tet* operator in the absence of doxycycline. This has been addressed by the development of Tet-repressible transcriptional repressors or suppressors tTR and tTS, which out-compete rtTA for *tet* operator sites in the absence of doxycycline, but are unable to bind when the antibiotic is present. They do not, however, physically interact with rtTA.

Numerous plasmid- and viral-based delivery vectors have been developed with tet-dependent transgene regulation for both *in vitro* and *in vivo* delivery, including retroviral, adenoviral and lentiviral systems. Many transgenic mice have been generated expressing Tet system components in specific

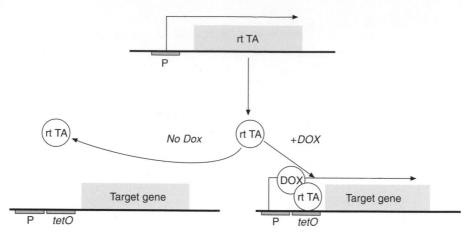

Figure 7.6

The reverse tet-activator system (rtTa), utilizing a mutated Tet repressor-VP16 fusion, which binds to DNA in the presence of tetracycline or doxycycline and releases DNA in its absence.

tissues, providing a ready source of donor strains to complement mice containing *tet*-regulated transgenes. Where multiple regulatory networks need to be studied concurrently, a second system is available based on the pristinamycin (Pip) repressor which is available both as repressible (PIT-Off) and inducible (PIT-On) versions regulatable by antibiotics of the streptogramin family. (See references (16–20).)

Hormone-dependent systems

The *Drosophila* ecdysone system is based on the steroid hormone ecdysone and its receptor, which orchestrate the extensive morphological changes that occur during molting in *Drosophila*. Since this hormone, its receptor and the DNA response elements recognized by the receptor are not found in mammalian cells, ecdysone and synthetic analogs such as muristerone A can be administered to mammalian cells and transgenic mammals without toxic effects. The system is not used in *Drosophila* itself.

The hormone acts by modulating the structure and assembly of a heterodimeric transcription factor formed from the *ecr* (*ecdysone receptor*) and *usp* (*ultraspiracle*) gene products. If an ecdysone response element is placed upstream of a transgene, then that transgene can be activated by exogenous muristerone A as long as the two *Drosophila* genes encoding the receptor are also expressed in the same cell. The first version of the ecdysone system used unmodified *Drosophila* gene products to assemble the transcription factor and the induction ratio was rather low. This situation was addressed by engineering the system to include mammalian components, producing a hybrid transcription factor recognizing a novel response element. For example, a modified ecdysone system has been described in which the *Drosophila* Ecr protein was fused to the herpes simplex virus VP16 transactivator and the mammalian glucocorticoid receptor, while the *usp* gene product was replaced by the retinoid X receptor. The novel receptor binds to ecdysone, recognizes a novel response element and has an induction ratio of $>10^3$.

Inducible systems based on mammalian steroid hormones have also been described. The most advanced system comprises a series of recombinant progesterone receptors carrying the DNA-binding domain of the yeast transcription factor GAL4, a truncated ligand-binding domain from the human progesterone receptor and the VP16 or NF-κB transactivation domains. These receptors are unable to bind progesterone itself, but can bind the progesterone antagonist (antiprogestin) RU486, also known as mifepristone. Transgenes driven by a promoter consisting of a TATA box and multiple GAL4 recognition sites are induced by RU486, apparently at levels low enough not to affect endogenous progesterone activity (*Figure 7.7*). Several versions of the system are available and have been developed commercially as the GeneSwitch platform by the biotechnology company Valentis Inc., which now licenses the technology to several major pharmaceutical and biotechnology companies (*Figure 7.8*). The systems have been used to regulate transgene expression in many transgenic *Drosophila* and mouse lines, and have been used in at least five human gene therapy trials to date. (See references (21,22).)

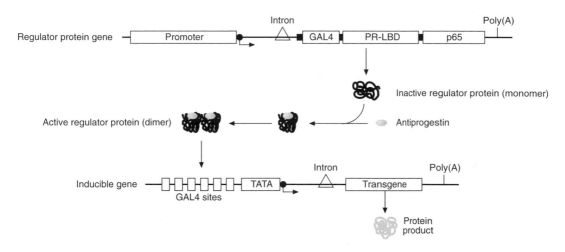

Figure 7.7

The GeneSwitch antiprogestin-inducible system. The antiprogestin-inducible regulator protein was constructed from the yeast GAL4 DNA-binding domain (GAL4), a truncated human progesterone receptor ligand-binding domain (PR-LBD), and an activation domain, such as the p65 subunit of human NF-κB (p65). The regulator protein is initially produced as an inactive monomer. Binding of the inducing drug (antiprogestin) triggers a conformational change that causes the regulator protein to become an activated homodimer, which binds to GAL4 sites in the inducible promoter, stimulating transcription of the transgene and leading to increased production of the protein product. When the antiprogestin is removed, the regulator protein reverts to its inactive, monomeric form and transcription of the transgene returns to baseline levels. Reprinted from Current Opinion in Biotechnology, Vol 13, No. 5, Nordstrom, "*Antiprogestin-controllable ...*", Pages 453–458, Copyright (2002), with permission from Elsevier.

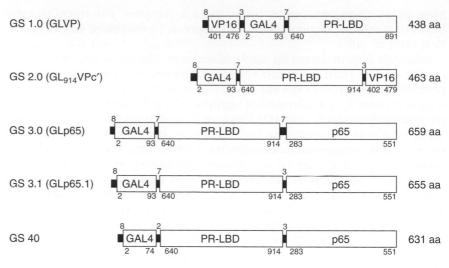

Figure 7.8

Versions of antiprogestin-inducible regulator proteins. GAL4 is a DNA-binding domain from the *Saccharomyces cerevisiae* GAL4 protein, PR-LBD is a C-terminal truncated version of the ligand-binding domain of the human progesterone receptor, VP16 is an activation domain from the herpes simplex virus VP16 protein and p65 is the activation domain from the p65 subunit of human transcription factor NF-κB. Numbers below individual segments indicate the boundaries of the selected amino acid residues and numbers above junctions indicate the lengths of linker regions. Reprinted from Current Opinion in Biotechnology, Vol 13, No. 5, Nordstrom, "*Antiprogestin-controllable ...*", Pages 453–458, Copyright (2002), with permission from Elsevier.

Chemically induced dimerization

Another system for inducible transgene regulation exploits the principles of chemically induced dimerization (CID), in which a divalent ligand simultaneously binds and thus brings together separate DNA-binding and transactivation domains to generate a hybrid transcription factor. The requirement for stable dimerization between the two components of the system minimizes background expression and thus provides a high induction ratio.

The original CID system exploited the immunosuppressant drug FK-506, which normally interacts with the single immunophilin domain of an immunoregulatory protein called FKBP (FK-506-binding protein). A divalent homodimer of this drug was synthesized. This synthetic molecule, named FK-1012, therefore had the ability to bind two immunophilin domains simultaneously. Hybrid components were constructed in which a DNA-binding domain and a transactivation domain were each fused to the immunophilin domain of FKBP. In the presence of FK-1012, the two components assembled into a heterodimeric transcription factor and activated the transgene.

A problem with this and other so-called *homodimerizer systems* is that the divalent ligand can also assemble homodimers comprising two DNA-binding domains or two transactivation domains. More efficient systems have therefore been produced in which heterodimeric ligands are used to avoid such non-productive interactions. An early system in which the immunophilin domain of FKBP was fused to the DNA-binding domain and cyclophilin (a cyclosporine A-binding domain) was fused to the transactivation domain is shown in *Figure 7.9*. Dimerization is stimulated by the synthetic molecule FKCsA, which is a composite of FK-506 and cyclosporine A. The most widely used dimerizer system is based on a naturally divalent molecule called rapamycin, which works by inducing the heterodimerization of FKBP and another protein called FRAP (FKBP-rapamycin-associated protein). Although rapamycin itself is not used *in vivo* because of potential immunosuppressive side effects, various analogs (rapalogs) are available that do not have significant pharmacological effects. Indeed, the best systems use rapalogs that no longer bind to endogenous FKBP and FRAP, but only interact with modified derivatives developed especially for the CID system. Other advantages of rapalogs include their prolonged bioavailability following oral administration and their ability to cross the blood–brain barrier. In the most popular version of the system (*Figure 7.10*) three copies of FKBP are joined to a synthetic DNA-binding domain comprising a pair of zinc fingers and a homeodomain, which binds a novel response element, whereas the FKBP-rapamycin binding domain of FRAP is joined to the p65

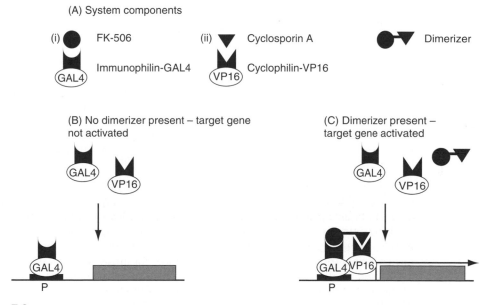

Figure 7.9

An early heterodimerizer system. (A) synthetic FK-506/cyclosporine dimerizer molecule or 'chemical inducer of dimerization' (CID) has the ability to recruit hybrid proteins comprising (i) the immunophilin domain of FKBP fused to the DNA-binding domain of the yeast transcription factor GAL4 and (ii) cyclophilin fused to the VP16 transactivator from herpes simplex virus. (B) In the absence of the dimerizer there is no expression. (C) the hybrid transcription factor assembles on a synthetic promoter (P) comprising multiple GAL4 binding sites and the target gene is activated. There is virtually no background transcription in the absence of CID.

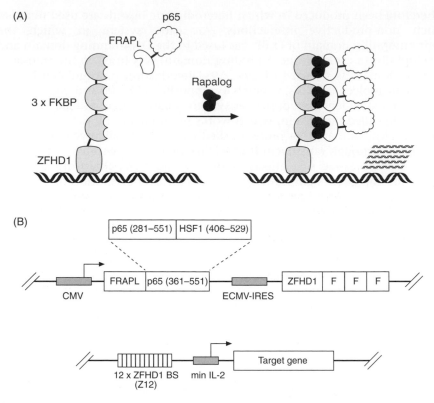

Figure 7.10

Configuration of the rapamycin system, the most commonly used dimerizer inducible gene expression system. (A) General scheme showing transcription factor fusion proteins. The DNA-binding domain fusion protein (gray) comprises the synthetic ZFHD1 DNA-binding domain fused to three copies of human FKBP. The activation domain fusion comprises the FRB domain of human FRAP (amino acids 2021–2113) fused to a portion of human NF-κB p65. The FRAP sequence incorporates the single point-mutation Thr2098Leu (FRAPL) to allow use of certain non-immunosuppressive rapamycin analogs. (B) Constructs for rapamycin-regulated expression. The transcription factor fusion proteins are expressed from the human cytomegalovirus (CMV) promoter using a bicistronic construct incorporating an IRES (internal ribosome entry site). The activation domain cassette is always placed before the IRES sequence to maximize its expression. Both fusion proteins are equipped with an N-terminal nuclear localization sequence from human c-Myc (not shown). The target promoter comprises 12 reiterated copies of the binding site for ZFHD1 (Z12) followed by a minimal promoter from the human interleukin-2 (IL-2) gene. Reprinted from Current Opinion in Biotechnology, Vol 13, No. 5, Pollack *et al*: "*Dimerizer-regulated gene . . .*", Pages 459–467, Copyright (2002), with permission from Elsevier.

activation domain subunit of NF-κB. Generally, both hybrid components are expressed from a single transcription unit using an internal ribosome entry site. The target promoter contains 12 copies of the target site followed by a TATA box. Other dimerizer systems are listed in *Table 7.1*. (See reference (23).)

Table 7.1 Summary of dimerizer systems used to regulate transcription

Dimerizer	Description	DNA-binding domain fusion	Activation domain fusion	Comments
Homodimerizers				
FK1012	FK506 homodimer	FKBP (3 copies)	FKBP (3 copies)	Does not work in stable cell lines
AP1510	Synthetic FKBP ligand homodimer	FKBP (3 copies)	FKBP (3 copies)	Works in stable cell lines
AP1889	'Bumped' analogs of AP 1510 series	FKBP with F36V mutation (3 copies)	FKBP with F36V mutation (3 copies)	Most potent homodimerizer system
Heterodimerizers				
Rapaymycin (and analogs)	Natural product (analog)	FKBP (1–3 copies)	FRB (with mutations(s), for analogs)	Most commonly used system
FKCsA	Dimer of FK506 and CsA	FKBP (3 copies)	Cyclophilin (2 copies)	Demonstrated in transient transfections
FK506 (and analogs)	Natural product (analog)	FKBP (1–3 copies)	Calcineurin A (CnA) or CAB (CnA–CnB fusion protein) (with mutation(s) for analogs)	Demonstrated in transient transfections in mammalian cells (also in yeast)
Dexa-FK506	Heterodimer of dexamethasone and FK506	Glucocorticoid receptor	FKBP	Used in three-hybrid studies in yeast
Dexa-MTX	Heterodimer of dexamethasone and methotrexate	DHFR	Glucocorticoid receptor	Used in three-hybrid studies in yeast
MTX-SLF	Heterodimer of methotrexate and a synthetic FKBP ligand	FKBP	DHFR	Used in three-hybrid studies in E. coli

Reprinted from Current Opinion in Biotechnology, Vol 13, No. 5, Pollack *et al*: *"Dimerizer-regulated gene . . ."*, Pages 459–467, Copyright (2002), with permission from Elsevier.

7.5 Transcriptional termination and RNA processing

During and after transcription in animal cells, the primary transcript is extensively processed before export from the nucleus for translation. The processing steps include the addition of an inverted 7-methylguanosine cap to the 5' end, the removal of introns by splicing, cleavage and polyadenylation to generate the mature 3' end of the transcript, and in some cases the methylation of certain internal bases. These processes must be considered carefully when designing expression vectors and their impact on transgene expression is discussed in more detail below.

Capping

The enzyme responsible for capping the 5' end of nascent transcripts in animal cells is intimately associated with the transcriptional initiation complex. Therefore, all eukaryotic mRNAs are capped directly after transcriptional initiation. The cap has several functions including protection from nucleases, facilitation of nuclear export and assembly of the ribosome prior to the initiation of protein synthesis. Most animal mRNAs possess what is known as a type 1 cap, which is methylated at position G^7 of the cap itself and at position $O^{2'}$ on the ribose moiety of the next residue. Since capping is an integral part of transcriptional initiation, no modules are required in expression vectors to facilitate this process.

Intron splicing

Introns are segments of DNA that are spliced out of the primary transcript and do not appear in the mRNA. The proportion of intron-containing genes in animal genomes appears to increase with the complexity of the animal. Therefore, in *Drosophila* and *Caenorhabditis*, about 80% of genes have introns and the average number of introns per gene is three, while in mammals and birds over 95% of genes have introns and the average number is about eight. About 10% of genes in mammalian genomes contain more than 30 introns. Introns are removed from the primary transcript by a well-characterized mechanism involving the recognition of 5' and 3' splice sites by a large, *trans*-acting assembly called the spliceosome. The activities of the spliceosome result in a *trans*-esterification reaction that joins the upstream and downstream exons and eliminates the intron (*Figure 7.11*).

Introns are significant for a number of reasons in expression vectors. First, it is notable that many vertebrate genes are extremely large but that most of the material within the gene is found in introns. An extreme example is

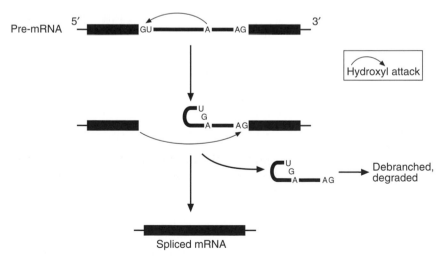

Figure 7.11

Cleavage of the 5' splice site is promoted by the hydroxyl (OH) attached to the 2'-carbon of an adenosine nucleotide within the intron sequence. This results in the lariat structure and is followed by the 3'-OH group of the upstream exon inducing cleavage of the 3' splice site. This enables the two exons to be ligated, with the released intron being debranched and degraded. Taken from Brown, T.A., *Genomes 2*, Copyright (2002), BIOS Scientific Publishers.

the human dystrophin gene *DMD1*, which is nearly 2 Mb in length but 97.5% of this is intron material and is eventually eliminated from the transcript. Given the limited capacity of many expression vectors used in animal cells, it is unsurprising that investigators often choose to express cDNA sequences, which are copied from mRNA and therefore lack introns, rather than complete genes. Alternatively, the introns can be removed artificially from genomic sequences to generate so-called minigenes, which are essentially the same in functional terms as cDNAs.

Although intron removal makes cloning and expression more practical and convenient, even the earliest cDNA expression studies provided evidence that splicing might be required for efficient transgene expression in mammalian cells. Many investigators have independently confirmed that cDNA expression is either intron-dependent or significantly improved in the presence of an intron, in some cases up to 500-fold. Therefore, most plasmid expression vectors now include a small heterologous intron, usually located within the 5′ UTR of the expression construct. Examples include the SV40 small t-antigen intron and the intron from the human growth hormone gene. In some vectors, hybrid introns or artificial introns have been inserted to match the consensus 5′ and 3′ splice sites. Although the presence of an intron has never been shown to inhibit transgene expression in plasmid vectors, it is equally clear that some genes do not require an intron for efficient expression. Indeed, 6% of mammalian genes lack introns even in their natural genomic context and introns are entirely absent from most bacterial genomes, yet bacterial genes such as *neo* and *lacZ* are efficiently expressed in mammalian cells.

Introns have not been shown to inhibit transgene expression when using plasmid vectors, but in certain viral transduction experiments the presence of an intron is very undesirable. As discussed in Chapter 4, the inclusion of sense-orientation introns in retroviral vectors can interfere with the viral replication cycle, and if an intron must be included in the expression construct, the entire transgene should be inverted and expressed from an opposing promoter so that the introns are not recognized during genomic transcription. In poxvirus vectors such as vaccinia virus, introns must be eliminated because the virus replicates in the cytosol and there is no splicing machinery available.

Why are introns so important and how do they influence gene expression? A possible post-transcriptional mechanism has been identified whereby primary transcripts are associated with different sets of proteins during synthesis, processing and export from the nucleus, depending on whether or not they contain introns. Some investigators have suggested that the absence of an intron can stimulate post-transcriptional silencing, presumably due to specific differences in the structure and components of the ribonucleoprotein particles into which all mRNAs are assembled. In *Xenopus* oocytes, an intron placed at the 5′ end of the transcript could alleviate this silencing, but a 3′ intron enhanced the repression. Similar results have been obtained in mammalian cells and introns in expression constructs are usually placed in the 5′ rather than the 3′ UTR.

In transgenic animals, particularly transgenic mammals, the presence of introns has been shown to be critical in many cases for reliable transgene expression. Indeed, full genomic constructs are much more strongly and faithfully expressed than cDNAs. The reasons for this phenomenon may include the presence of long-distance regulatory elements within and beyond the transcription unit, which are necessary to maintain an open chromatin domain in the context of a transgenic animal. It must be stressed that cultured cells are highly artificial systems where the normal rules of chromatin organization and chromatin-mediated gene regulation may not be so stringent (Chapter 8). Furthermore, cells in which transgenes are silenced by chromatin position effects simply fail to survive selection, so there is a bias towards those cells where the transgene has integrated at a permissive site. With transgenic animals there is a much smaller sample number, and more careful consideration must be given to transgene design.

Another consideration involving introns is the generic nature of the splice signals, often GU at the 5' splice site and AG at the 3' splice site. Such signals can easily be introduced adventitiously into a transgene, resulting in so-called cryptic splice sites that will generate aberrant products. Despite this potential problem, eukaryotic primary transcripts are generally spliced in a predictable fashion even though all introns carry the same consensus splice signals, which are short enough to occur by chance both within the body of the intron and in adjacent exons. It is thought that the sequence context of the splice sites, the distance between them, and the secondary structure adopted by the nascent transcript or completed primary transcript may dictate how the splicing apparatus and RNA interact to define the introns to be removed.

Transcriptional termination and polyadenylation

The termination of RNA polymerase II transcription occurs downstream of the mature 3' end of the transcript, and may even be a random event. The mature end of the transcript is generated by endonucleolytic cleavage and this is followed by a process known as polyadenylation, which involves the addition of approximately 200 adenylate residues to generate the poly(A) tail. The precise function of polyadenylation is unclear, but it may increase mRNA stability and regulate mRNA export from the nucleus. Polyadenylation takes place 10–30 nucleotides downstream of a highly conserved polyadenylation site, with the consensus sequence AAUAAA. Cleavage and polyadenylation are carried out by a multimeric complex comprising recognition factors that bind to the polyadenylation site, a dimeric cleavage enzyme and the enzyme polyadenylate polymerase (*Figure 7.12*).

Like promoters and enhancers, different polyadenylation sites vary in their 'strength' and can influence transgene expression to different degrees. Some of the most active sites are included in contemporary expression vectors, e.g. those from the SV40 early transcription unit, and the mouse β-globin gene. Experiments have shown that transgenes including such polyadenylation sites are up to ten times more active than those without.

Polyadenylation sites are particularly important if the expression vector contains two or more independent transcription units, since the polyadenylation sites will help to prevent transcription from one transgene running through into the others. Without this constraint, transcription from one transgene could run through into the downstream transgene and occlude

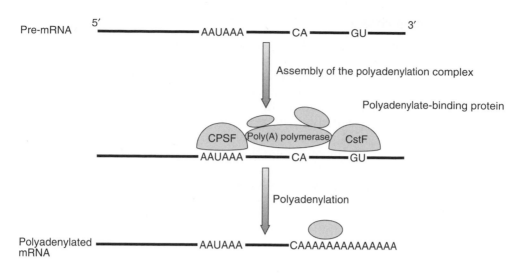

Figure 7.12

Polyadenylation of eukaryotic mRNA. Taken from Brown, T.A., *Genomes 2*, Copyright (2002), BIOS Scientific Publishers.

the promoter, preventing the assembly of the transcription complex and blocking gene expression. If adjacent genes are arranged as inverted repeats, read-through transcription from one transgene can generate antisense RNA from the other, resulting in antisense suppression (Chapter 9). If sense and antisense RNA are generated from the same transgene, the resulting dsRNA can induce a general interferon response which blocks global protein synthesis, and may even result in RNA interference (Chapter 9), causing potent and long-lasting silencing of the transgene. (See references (24,25).)

7.6   Protein synthesis

Kozak consensus

The optimal translational start site sequence in animal cells was defined by Kozak as 5′ CCRCCAUGG 3′, with the underlined sequence representing the initiation codon (R is any purine). The two most important residues appear to be the purine at position –3 and the guanosine at position +4 relative to the start site[2]. Most expression vectors contain the optimized Kozak consensus sequence as standard, although only proteins with valine, alanine, glutamine, glycine or aspartate as their second residue can be specified by a coding region with guanosine at position +4. Mutations deviating from the Kozak consensus can reduce protein expression levels by up to 90%, even if the AUG codon is retained.

---

2 Note that by convention there is no position 0 in a gene. The translation start is defined as position +1 and is usually the A of the AUG codon. The nucleotide immediately preceding this is defined as position –1.

Codon choice

Since the genetic code is degenerate, most amino acids are specified by more than one codon and some are specified by as many as six. Different organisms have different preferences for codon usage within these degenerate families, and this phenomenon is known as *codon bias*. This is of interest because the codon bias in the donor of a transgene may differ from that in the host, particularly if the two organisms are dissimilar (e.g. bacterial or yeast transgenes expressed in mammalian cells).

There are several principles that underlie codon bias, including the differential abundance of particular iso-accepting tRNAs[3] in different species, variations in the 'wobble rules' affecting third base codon/first base anticodon pairing and different genomic base compositions, which means that codons with a GC bias may be favored over those with an AT bias or vice versa. Whatever the reasons for differences, the practical limitations generally reflect inefficient protein synthesis when rare codons are encountered by the host. In many species, rare codons (i.e. codons that are least favored in a particular degenerate family) are used as a regulatory mechanism, and will result in pausing (delay before amino acid incorporation), skipping (jumping over the codon to the next one), misincorporation (inserting the wrong amino acid) or even frameshifting (slipping into an alternative frame) or truncation (abortion of protein synthesis). These problems can be addressed by codon optimization, bringing the codon usage in the transgene in line with that of the host. This is achieved by introducing translationally neutral mutations into the coding region. Occasionally, codon optimization may have unpredictable effects, which result in the loss of transgene expression – mutations that do not affect the coding sense of the transcript may nevertheless influence mRNA stability or introduce an unusual context. For example, the termination codon UGA can, in the correct context, be interpreted as sense codon leading to the insertion of the rare amino acid selenocysteine.

Structure of non-coding regions

The non-coding regions of the transgene are transcribed but not translated, however their structure can influence the efficiency of transgene expression by affecting mRNA stability and/or the efficiency of protein synthesis. The structure of the 5′ untranslated can influence ribosome assembly and scanning. The small ribosome subunit attaches at the 5′ 7-methylguanosine cap, and scans along the mRNA until the first AUG codon is encountered. If the 5′ UTR contains a significant amount of secondary structure, assembly and scanning can be inhibited. The presence of multiple AUG codons upstream of the genuine start site is also deleterious, especially if one or more of them are followed by in-frame termination codons. Most expression vectors are therefore designed to incorporate only a minimal 5′ UTR sequence, although as discussed above there may be a small intron present, which is spliced out of the primary transcript.

---

3 These are tRNAs that carry the same amino acid but have different anticodon sequences and therefore recognize different codons.

The 3′ untranslated region of the native gene is usually eliminated from the expression construct and the transgene contains only a minimal sequence. This is because native 3′ UTRs often harbor regulatory elements that control mRNA stability and protein synthesis.

Internal ribosome entry sites

Internal ribosome entry sites (IRES) are unusual elements that were once thought to be present only in certain viral genomes but are now also known to occur in a small number of endogenous genes in animal cells. As their name suggests, they allow the assembly of ribosomes in the middle of a messenger RNA molecule, instead of at the 5′ cap, which is usually mandatory for ribosome assembly. IRES elements therefore provide a very useful mechanism for the expression of multiple transgenes, in which two or more genes can be expressed under the control of a single promoter. The transgenes are transcribed from this promoter to yield a single, polycistronic message, but they are translated independently.

IRES elements have many uses in gene transfer experiments, several of which involve the linked co-expression of a primary transgene and a selectable marker or reporter gene. For example, a reporter gene such as *lacZ* or *gfp* can be placed downstream of an IRES in the same transcription unit as a primary transgene. This allows cells expressing the foreign gene to be identified through co-expression of the reporter, even if the primary transgene does not generate a readily identifiable phenotype. Another example is the expression of a reporter gene in the targeted locus of a knockout mouse. The insertion of an IRES-*lacZ* construct into an endogenous gene facilitates both knockout by insertional mutagenesis and the identification of cells which now lack the targeted gene product. The expression of genes downstream of an IRES is usually much less efficient than normal 5′ cap-dependent expression, but is generally sufficient for marker gene expression. (See references (26–29).)

7.7 Post-translational events

Overview

After synthesis, all polypeptides are modified in some fashion. Such modifications may include cleavage of the polypeptide backbone, minor chemical changes to individual amino acid side chains and the addition of bulky chemical groups such as aryl chains or complex oligosaccharides. Some forms of protein modification are reversible, and can be used as a means to regulate protein activity (e.g. phosphorylation of serine, threonine and tyrosine residues), while others are permanent. Protein modification may be necessary for structural maturation or catalytic activity, it may facilitate interactions with other molecules, or it may be required for correct trafficking to particular compartments within the cell. The relationship between protein trafficking and modification is intimate, as certain modifications precede sorting, and sorting to certain compartments is required for certain forms of modification. For example, cleavage of the signal peptide is necessary for proteins to be admitted into a secretory pathway, which is in turn necessary for those proteins to be glycosylated (*N*-linked glycosylation occurs in the ER lumen, and *O*-linked glycosylation occurs further downstream in the Golgi network). This in turn may be necessary for further sorting (the addition of mannose-6-phosphate groups targets proteins to lysosomes). Furthermore, some forms of protein modification are involved in the regulation of protein degradation (e.g. ubiquitinylation of lysine residues).

The secretion of proteins from mammalian cells involves a characteristic series of modifications. Secreted proteins possess an N-terminal signal peptide that causes the ribosome to associate with the ER membrane. The protein is then cotranslationally translocated into the ER lumen, where the signal peptide is cleaved and a preformed high-mannose oligosaccharide is added to asparagine residues within the consensus sequence Asn-Xaa-Ser/Thr (where Xaa is any residue except proline). Several terminal residues may be removed from this basic *N*-linked glycan in the ER before the immature glycoprotein is transported to the Golgi network, where further modifications take place. The Golgi is also the compartment where *O*-linked glycans may be added to serine and threonine residues. As well as glycosylation, acylation and other forms of modification also occur in the secretory pathway, e.g. the O^4-sulfation of tyrosine residues, and various forms of proteolytic processing.

Factors affecting the processing of recombinant proteins in animal cells

The processing of recombinant proteins in animal cells is governed by the targeting information included in the expression construct as well as intrinsic properties of the protein itself and the qualities of the host cell. For example, a mammalian protein that is normally glycosylated in its native setting might not be glycosylated in its heterologous setting for the following reasons:

- If the native signal sequence has been trimmed off during cloning, and the expression vector lacks a signal sequence (properties of the expression construct). This problem needs to be addressed especially if only a partial cDNA sequence is available for translation, or if a recombinant protein is to be expressed as an N-terminal fusion protein, because the endogenous signal peptide would either be missing or obscured.
- If the glycosylation target site on the proteins has been abolished by alterations to the transgene sequence (intrinsic properties of the protein).
- If the host cell lacks the appropriate enzymes to complete normal glycosylation (properties of the host).

The modification of a protein is also dependent on the sequence context surrounding the modification motif and the overall tertiary conformation of the protein. Both these factors may influence how the protein is perceived by the modification enzymes in the host cell. Where modification occurs in a series of successive steps, as in glycosylation, each modification may help to make the protein accessible to the next processing enzyme. Therefore even subtle changes in sequence may interfere with this process and prevent subsequent modification steps. This is demonstrated by the alteration of, e.g. glycosylation patterns following mutations causing non-conservative amino acid changes, or following the incorporation of amino acid analogs. The dependence of late processing events on the successful completion of earlier modifications is demonstrated by the altered glycosylation patterns observed following treatment with inhibitors of the early glycosylation enzymes.

The qualities of the host cell play an important role in the synthesis of recombinant proteins and this is most noticeable when taxonomically different transgene donors and expression hosts are used. For example, the expression of mammalian transgenes in insect cell lines derived from *Spodoptera frugiperda* (e.g. Sf9) generally results in the production of recombinant proteins with glycan structures very different to those of the native proteins. This is because *S. frugiperda* lacks a key gene present in mammals, encoding the enzyme β-1,2-N-acetylglucosaminyltransferase. Thus, proteins expressed in Sf9 cells do not produce certain glycosylation intermediates, such as side chains ending in galactose-sialic acid residues. Other insects do possess this enzyme, so alternative cell lines can be used to produce authentic glycoproteins. It has also been possible to co-express a primary transgene with the gene encoding β-1,2-N-acetylglucosaminyl transferase in baculovirus vectors in Sf9 cells. Some differences in glycosylation patterns have even been observed between alternative mammalian hosts, e.g. secreted proteins produced in CHO cells tend to have more terminal sialic acid residues than those produced in COS cells, so this may need to be taken into account when choosing a particular host–vector system.

The overexpression of proteins that require proteolytic cleavage as part of their processing can also present difficulties, as the endogenous proteases, *proprotein convertases*, are often present at low levels. A number of groups have successfully co-expressed the precursor and its cognate proprotein convertase in an attempt to generate high yields of mature recombinant protein, e.g. insulin in insect cells, von Willebrand factor in CHO cells and coagulation factor C in transgenic pigs. (See references (30–37).)

7.8   Inducible protein activity

The inducible transcription systems described earlier in this chapter allow transgene activity in cells and transgenic animals to be controlled by the application of a small molecule (an inducer) such as IPTG, doxycycline, a synthetic hormone analog or an inducer of dimerization. The problem with all such systems is the delay inherent between the activation of transcription and the response at the level of recombinant protein expression. There is a lag of several hours before induction peaks, and there is a similar delay in the return to the basal state when induction is removed.

In order to avoid this lag, inducible expression systems have also been developed which work at the post-transcriptional level, one involving a steroid-binding fusion protein and one involving chemically induced dimerization. The steroid-based method exploits the tendency of the estrogen receptor to interact with a protein called heat shock protein 90 (Hsp90) in the absence of estrogen to form an inactive complex. When estrogen is present, it causes a conformational change in the receptor, which then dissociates from Hsp90 and binds to its response element in the DNA. The

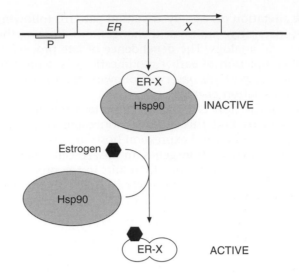

Figure 7.13

The estrogen-inducible expression system. A protein (X) expressed as a fusion with the estrogen receptor (ER) is generally inactive because it is sequestered into a complex with heat shock protein 90 (Hsp90). In the presence of estrogen, however, the fusion protein is released from the complex and the activity of protein X is restored. This is an advantageous system because induction is very rapid, requiring only the dissociation of a protein complex and not transcription followed by protein synthesis.

principle of the estrogen-based regulation system is that any protein expressed as a fusion with a hormone-binding domain of the estrogen receptor should also form an inert complex with Hsp90, but will similarly be released from repression in the presence of estrogen. Therefore, a protein can be expressed as a fusion product under the control of a constitutive promoter in a cell line or transgenic animal and will accumulate as an inert product, but will be activated in seconds, i.e. without an 'expression lag', if estrogen is added to the culture medium or food/water (*Figure 7.13*). A more recently developed version of this system uses a modified estrogen receptor, which is highly specific for the ligand 4-hydroxytamoxifen, thus removing many of the undesirable effects of estrogen administration.

The chemically induced dimerization (CID) system takes advantage of the ability of ligands to facilitate the dimerization of proteins carrying ligand-binding domains. The CID system has already been described as a transcriptional switch, in which separate DNA-binding and transactivation domains expressed as ligand-binding fusion proteins are conditionally dimerized by supplying a divalent ligand. The same idea can be exploited to activate proteins whose normal activation requires a dimerization step (e.g. many receptors and signal transduction proteins) or the separately expressed domains of monomeric proteins (as for transcription factors in the transcriptional CID switch). CID has been successfully used to regulate the activity of a range of transmembrane receptors and intracellular signaling proteins by conditional activation. (See references (38–40).)

References

1. Sambrook J, Fritsch EF, Maniatis T (1989) Expression of proteins. In: Sambrook J, Fritsch EF, Maniatis T (eds) *Molecular Cloning – A Laboratory Manual* (second edition), pp. 16.3–16.29. CSH Press, Cold Spring Harbor, NY.
2. Chisholm V (1995) High efficiency gene transfer into mammalian cells. In: Glover DM, Hames BD (eds) *DNA Cloning – A Practical Approach Volume 4 Mammalian Systems* (second edition), pp. 1–42. IRL Press, Oxford, UK.
3. Latchman DS (2000) *Eukaryotic Transcription Factors* (third edition). Academic Press, London, UK.
4. Latchman DS (1995) *Gene Regulation, A Eukaryotic Perspective* (second edition). Chapman & Hall, New York.
5. Mulligan RC, Berg P (1981) Selection for animal-cells that express the Escherichia-coli gene coding for xanthine-guanine phosphoribosyl transferase. *Proc Natl Acad Sci USA* **78**: 2072–2076.
6. Gorman CM, Merlino GT, Willingham MC (1982) The rous-sarcoma virus long terminal repeat is a strong promoter when introduced into a variety of eukaryotic cells by DNA-mediated transfection. *Proc Natl Acad Sci USA* **79**: 6777–6781.
7. Boshart M, Weber F, Jahn G, Dorsch-Hasler K, Fleckenstein B, Schaffner W (1985) A very strong enhancer is located upstream of an immediate early gene of human cytomegalo-virus. *Cell* **41**: 521–530.
8. Gorman CM, Gies D, McCray G, Huang M (1989) The human cytomegalo-virus major immediate early promoter can be trans-activated by adenovirus early proteins. *Virology* **171**: 377–385.
9. Wall RJ (1999) Biotechnology for the production of modified and innovative animal products: transgeneic livestock bioreactors. *Livestock Prod Sci* **59**: 243–255.
10. Wurm FM, Gwinn KA, Kingston RE (1986) Inducible overproduction of the mouse c-Myc protein in mammalian cells. *Proc Natl Acad Sci USA* **83**: 5414–5418.
11. Bendig MM, Stephens PE, Crocket MI, Hentschel CC (1987) Mouse cell lines that use heat shock promoters to regulate the expression of tissue plasminogen activator. *DNA* **6**: 343–352.
12. Lee F, Mulligan R, Berg P, Ringold G (1981) Glucocorticoids regulate expression of dihydrofolate-reductase CDNA in mouse mammary-tumor virus chimaeric plasmids. *Nature* **294**: 228–232.
13. Hager LJ, Palmiter RD (1981) Transcriptional regulation of mouse liver metallothionein I gene by glucocorticoids. *Nature* **291**: 340–342.
14. McCormick F, Trahey M, Innis M, Dieckmann B, Ringold G (1984) **(Q1)** *Mol Cell Biol* **4**: 166–172.
15. Saez E, No D, West A, Evans RM (1997) Inducible gene expression in mammalian cells and transgenic mice. *Curr Opin Biotechnol* **8**: 608–616.
16. Hu MC, Davidson N (1987) The inducible IAC operator-repressor system is functional in mammalian-cells. *Cell* **48**: 555–566.
17. Goosen M, Bonin AL, Bujard H (1993) Control of gene activity in higher eukaryotic cells by prokaryotic regulatory elements. *Trends Biochem Sci* **18**: 471–475.
18. Shockett PE, Schatz DG (1996) Diverse strategies for tetracycline-regulated inducible gene expression. *Proc Natl Acad Sci USA* **93**: 5173–5176.
19. Corbel SY, Ross FMV (2002) Latest developments and in vivo use of the Tet sysmte: ex vivo and in vivo delivery of tetracycline-regulated genes. *Curr Opin Biotechnol* **13**: 448–452.
20. Baron U, Gossen M, Bujard H (1997) Tetracycline-controlled transcription in eukaryotes: Novel transactivators with graded transactivation potential. *Nucleic Acids Res* **25**(14): 2723–2729.
21. No D, Yo TP, Evans RM (1996) Ecdysone-inducible gene expression in mammalian cells and transgenic mice. *Proc Natl Acad Sci USA* **93**: 3346–3351.

22. Nordstrom JL (2002) Antiprogestin-controllable transgene regulation in vivo. *Curr Opin Biotechnol* **13**: 453–458.

23. Pollock R, Clackson T (2002) Dimerizer-regulated gene expression. *Curr Opin Biotechnol* **13**: 459–467.

24. Kaufman RJ (1990) Vectors used for expression in mammalian-cells. *Methods Enzymol* **185**: 487–511.

25. Palmiter RD, Sandgren EP, Avarbock MR, Allen DD, Brinster RL (1991) Heterologous introns can enhance expression of transgenes in mice. *Proc Natl Acad Sci USA* **88**: 478–482.

26. Kozak M (1986) Point mutations define a sequence flanking the AUG initiator codon that modulates translation by eukaryotic ribosomes. *Cell* **44**: 283–292.

27. Kozak M (1984) Selection of initiation sites by eukaryotic ribosomes – effect of inserting aug triplets upstream from the coding sequence for preproinsulin. *Nucleic Acids Res* **12**: 3873–3893.

28. Shaw G, Kamen R (1986) A conserved AU sequence from the 3′ untranslated region of GM-CSF messenger –RNA mediates selective messenger –RNA degradation. *Cell* **46**: 659–667.

29. Mountford PS, Smith AG (1995) Internal ribosome entry sites and dicistronic RNAs in mammalian transgenesis. *Trends Genet* **11**: 179–184.

30. Kornfield R, Kornfield S (1985) **(Q1)** In: Lennarz W (ed.) *The Biochemistry of Glycoproteins and Proteoglycans*, pp. 34–87. Plenum Press, New York.

31. Sakaguchi M (1997) Euksryotic protein secretion. *Curr Opin Biotechnol* **8**: 595–601.

32. Dorner AJ, Bole DJ, Kaufman RJ (1987) The relationship of N-linked glycosylation and heavy-chain binding-protein association with the secretion of glycoproteins. *J Cell Biol* **105**: 2665–2674.

33. Hsieh P, Rosner MR, Robbins PW (1983) Host-dependent variation of asparagines-linked oligosaccharides at individual glycosylation sites of sindbis virus glycoproteins. *J Biol Chem* **257**: 2548–2554.

34. Williams DB, Lennarz WJ (1984) Control of asaparagine linked oligosaccharide chain processing: studies on bovine pancreatic ribonuclease B. An *in vitro* system for the processing of exogenous glycoprotiens. *J Biol Chem* **259**: 5105–5114.

35. Davies AH (1995) Advances in the use of recombinant baculoviruses for the expression of heterologous proteins. *Curr Opin Biotechnol* **6**: 543–547.

36. Wagner R, Liedtke S, Kretzschmar E, Geyer H, Geyer R, Klenk H-D (1996) Elongation of the N-glycans of fowl plague virus hem agglutinin expressed in Spodoptera frugiperda (Sf9) cells by coexpression of human beta 1, 2–N-acetyl-gluosaminyltransferase I. *Glycobiology* **6**: 165–175.

37. Seidah NG, Chretien M (1997) Eukaryotic protein processing: endoproteolysis of precursor proteins. *Curr Opin Biotechnol* **8**: 602–607.

38. Picard D (1994) Regulation of protein function through expression of chimaeric proteins. *Curr Opin Biotechnol* **5**: 511–515.

39. Porter A (1998) Controlling your losses: conditional gene silencing in mammals. *Trends Genet* **14**: 73–79.

40. Spencer DM (1996) Creating conditional mutations in mammals. *Trends Genet* **12**: 181–187.

Transgene behavior

8

8.1 Introduction

In Chapter 7 we considered different aspects of vector design that can be used to control transgene expression. The salient point here is that vector design allows transgene expression to be controlled by the investigator, and the effects of vector modification are usually predictable. However, there are also many influences on transgene expression which are unpredictable, and most of these occur inside the cell where the investigator has little influence. As stated at the beginning of this book, the integration of exogenous DNA by illegitimate recombination is regarded as an approximately random process depending to a large degree on the presence of pre-existing chromosome breaks[1]. We say *approximately* random because there does appear to be a certain bias for insertion in all genomes, probably reflecting the normal state of chromatin architecture and its accessibility to foreign DNA. With transpositional vectors such as P-elements and retroviruses, there may even be a degree of target site choice at least at the sequence level, although the consensus target sequences are generally weak with the result that insertions viewed on a genomic scale remain more or less random. The position of integration is therefore one aspect of gene transfer than cannot be controlled unless specialized gene-targeting strategies are used (Chapter 6). Another is the structure of the resulting transgenic locus – the number of transgene copies, their arrangement, their structural integrity – all of these lie outside the control of the investigator again unless gene targeting is used to introduce precise modifications.

What are the consequences of these uncontrollable events? A frequent observation made in gene transfer experiments is that transgene expression is highly variable among cell lines or transgenic animals generated by independent transformation events. In addition, transgene integration may occasionally cause unexpected changes in phenotype resulting from serendipitous rearrangements of the host genome. Conversely, transgene expression within a clone of transformed cells, or within a line of transgenic animals, is much more likely to be comparable, although this is not always the case.

Names have been given to the uncontrollable aspects of gene transfer so that variable transgene expression or loss of transgene expression can be blamed on specific phenomena. The variable position of transgene integration gives rise to *position effects* that reflect the influence of the local molecular environment on transgene expression, while the variable

---

1 In plants, it has been suggested that particle bombardment may *cause* some chromosome breaks as the metal particle skims the ends off chromatin loops in the nucleus, and therefore pre-existing breaks are not required. The situation in animal cells is not clear but the same principles could well apply.

number of integrated transgene copies gives rise to *dosage effects* that may in some cases reflect the increase in protein levels concomitant with transgene copy number. The variable structure and organization of transgene loci can lead to a number of related *gene-silencing* phenomena including the production of aberrant RNA species and interactions between DNA sequences to form secondary structures. Finally, *integration effects* can abolish, modify or even enhance the expression of genes at or near the integration site. All these factors may conspire to produce highly variable and inconsistent results from gene transfer experiments with essentially the same transformation vector.

8.2 Timing of transgene integration

Before considering differences that may occur between clones of cells or individual animals transformed *separately* with the same exogenous DNA, it is worth remembering that even cells within the same clone or the same animal can behave differently in response to exogenous DNA, resulting in extensive mosaicism for transgene integration and expression.

When non-replicative DNA is introduced into cultured cells by transfection, individual cells are initially selected on the basis of their ability to express a marker gene over a period of weeks, which is *prime facie* evidence for stable integration. However, if selection is removed, individual cells within a clone can lose the transgene, for it is no longer required for survival, and over time the non-transformed cells usually out-compete their transformed cousins. Conversely, if the intensity of selection is stepped up, certain marker genes allow the selection of cells with massively amplified transgene arrays (page 16). It is important to remember that the selection regime does not induce the amplification process but merely selects for amplifications that have occurred randomly. In the absence of selection, the highly amplified arrays would still form, but would offer no selective advantage so the cells containing them would not prosper. The take home point is that genetic variation at the transgene level occurs naturally, and population homogeneity is only maintained by continuous selection.

In transgenic animals the situation is different because there are millions to trillions of cells, none of which are maintained under selection. Differences at the cellular level are therefore bound to occur, although most such changes are unlikely to be noticed in the context of the entire animal. An exception, however, is the occurrence of such transgene modifications in early development, resulting in noticeable mosaicism at the whole animal level. This can occur in two ways: a delay in transgene integration, or a spontaneous instance of transgene excision. Both are fairly common events, which is why germline transmission is required for proof of transgenesis.

Taking the mouse as an example, DNA introduced into the fertilized egg by pronuclear microinjection often integrates after one round of DNA replication, which results in the exclusion of the transgene from one daughter cell in the zygote[2]. If the DNA fails to integrate prior to the first mitotic division, then several outcomes are possible. If episomal DNA persists in both daughter nuclei, the transgene can integrate at two independent sites in the two daughter cells, or in only one, or neither of the cells. If integration occurs in one daughter cell before the next round of replication, the transgene will be present in all descendants of that cell, but if another round of replication occurs, the transgene will be excluded from three of the four cells. The status of the resulting mouse then depends on which cells give rise to the inner cell mass and which to the trophoblast. None of these events can be controlled.

Mice, and mammals generally, undergo a characteristic form of development where very little material is stored in the egg, and zygotic transcription begins as early as the two-cell stage. Conversely, animals such as fish and amphibians have large eggs with a stockpile of proteins and RNA. The early embryo undergoes a rapid series of synchronous cleavage divisions followed by an event termed the midblastula transition, where division becomes asynchronous and zygotic transcription begins. This event is thought to reflect the titration of DNA packaging proteins such as histones. It is the initial surplus of these DNA packaging proteins that may be responsible for the unique behavior of DNA introduced into amphibian eggs. Unlike mammalian systems, where the DNA is soon integrated, exogenous DNA in frog eggs is rapidly concatemerized and subject to extrachromosomal replication. As the embryo cleaves, the distribution of transgene copies is arbitrary, resulting in mosaic and variable expression levels between equivalent cells. Eventually, the exogenous DNA is degraded, but at some point integration may occur. While this is an early event in mouse development, with frogs, integration often occurs much later, and the resulting mosaics may contain only a low percentage of transformed cells (reflecting integration after e.g. 12 rounds of cleavage). (See references (1,2).)

8.3 The fate of foreign DNA and the structure of the transgenome

Exogenous DNA reaching the nucleus of an animal cell rapidly becomes covalently joined to form larger arrays, whose structure is dependent on a number of variables including the amount and conformation of the DNA, and its complexity. It is most likely that the arrays form prior to integration although a mechanism in which an individual integrated transgene acts as a hotspot for further integration events cannot be excluded. In calcium phosphate transfection, the cell is presented with a large amount of DNA which may include multiple plasmids and fragments of carrier DNA. At some point in the transformation process, these sequences become linked into large structures that have been described as 'transgenomes', and which comprise mixed arrays of the various transgenes and vector backbone sequences interspersed with fragments of carrier DNA. These structures may be 2–3 Mb in length. Integration tends to occur at a single

---

2 Note that in mammals, the male and female pronuclei undergo a round of replication before they fuse together and then undergo the first round of mitosis. If the transgene integrates prior to replication, it will be present in both copies of the male pronuclear genome and will pass to both daughter cells of the zygote. If integration occurs after replication, it will be present only on one copy of the genome and will pass to only one daughter cell.

site and can be unstable, i.e. subject to full or partial excision probably reflecting intrachromosomal recombination events. In other transfection methods, carrier DNA is not required and the amount of DNA entering the cell is lower. The DNA is also less complex. Arrays are still generated but they are much smaller than the transgenome described above, and are often organized as simple direct repeats. The reason for this may reflect the impact of exogenous DNA on the DNA repair systems of the cell. In calcium phosphate transfection, the host cell's nucleus becomes swamped with DNA fragments and the relative concentration of DNA ends is much higher than normal. This may stimulate the production of DNA ligase, resulting in extensive concatemerization. Linearization of the vector improves the efficiency of stable transformation in most transfection methods, probably by stimulating DNA ligase activity. However, as discussed in Chapter 6, linearization also improves the efficiency of homologous recombination so it is likely that multiple recombination mechanisms are involved in the formation of transgene arrays. (See references (3–6).)

8.4 Position effects

In the context of gene transfer experiments, position effects are defined as influences on transgene activity brought about by environmental differences between different integration sites. They are a major cause of differential transgene expression levels among independently derived transformed cell lines or transgenic animals, and they are also thought to reduce or eliminate copy-number dependence in transgene expression. The position of integration can influence transgene expression through at least three distinct mechanisms (*Figure 8.1*): the activity of local regulatory elements, the local chromatin structure, and the local state of DNA methylation. These factors are interdependent since condensed chromatin is generally hypermethylated, and active regulatory elements are generally associated with relaxed and non-methylated chromatin.

As a general example of how position effects work, consider an experiment carried out by Bishop and colleagues to determine the activity of an anomalous transgene (7). In this experiment, a series of transgenic mice was produced using a single expression construct, but one of the mice showed anomalous expression of the transgene. The transgene itself appeared to be intact, so the investigators recloned the transgene from the genomic DNA of the transgenic mouse and used it to create another transgenic line by pronuclear microinjection. The secondary transgenic mice expressed the transgene as expected, showing that the anomalous expression profile in the original mice was not caused by structural modification of the transgene, but by an epigenetic effect.

The most specific form of position effect occurs when an integrated transgene falls under the influence of a local regulatory element. For example, if a transgene integrates within range of a nearby enhancer, the transgene may come under the control of that enhancer and may partially replicate the expression profile it specifies. Transgenes with weak promoters and enhancers are the most susceptible to such position effects, but local enhancers (and silencers) will affect any transgene construct, even those with strong and specific promoters and enhancers of their own.

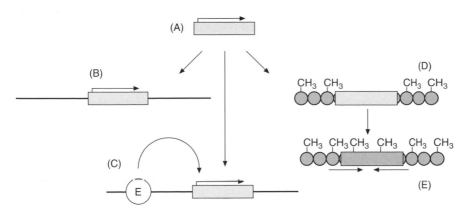

Figure 8.1

Position effects influencing transgene expression. (A) A non-integrated transgene may be expressed strongly. (B) Upon integration, the transgene may be expressed as expected, suggesting it has integrated into an active region of the genome. (C) However, it may come under the influence of an endogenous enhancer or other local regulatory elements, which may alter the expected expression pattern. (D) The transgene could also integrate into repressed and hypermethylated chromatin, resulting in reduced expression or silencing. (E) The repressed chromatin may spread into the transgene, eliminating its expression all together.

Independently derived transgenic animals carrying the same expression construct often show a 'core' expression pattern reflecting the activity of the regulatory elements included in the vector, but this may be supplemented with unique patterns in some lines reflecting ectopic activation of the transgene by local enhancers, and there may be incomplete expression in other lines reflecting the presence of specific silencers. This form of position effect has been widely exploited in the development of the *enhancer trap*, an insertional vector containing a very weak promoter that relies entirely on the presence of local enhancers for its activation (Chapter 9).

Transgenes can also integrate within endogenous genes, in which case they may be influenced by local promoter elements or splice signals. Again, there are entrapment vectors that exploit this phenomenon, and these are known as *gene traps*. As discussed in Chapter 9, such vectors contain either a 'naked' reporter gene, i.e. one driven simply by a translational start codon that can respond to adjacent promoter elements, or a splice acceptor site that enables the vector to function as an ectopic exon when it integrates into an endogenous intron. Transgenes integrating at the 5′ end of a gene may come under the control of the endogenous promoter, or the endogenous and transgene promoter elements may combine in a novel and unexpected manner. Transgenes that integrate into the body of an endogenous gene may be expressed either from their own promoter, or as an aberrant hybrid mRNA under the control of the endogenous gene promoter.

Not all position effects are as specific as those described above. Gene expression is dependent not only on regulatory elements with specific qualitative effects, but also on the local chromatin structure which has a more general, quantitative influence. Chromatin is often described in terms of two states, one of which is 'open' and transcriptionally active, and the other of which is 'closed' and transcriptionally repressed. Open and closed chromatin have a number of contrasting properties such as different levels of histone acetylation and DNA methylation, differential distribution of particular DNA-binding proteins and differential nuclease sensitivity. A transgene integrating into inactive chromatin (heterochromatin) is often subject to position-dependent silencing as the repressed chromatin can spread from the surrounding sequences into the transgene and abolish its expression. In contrast, a transgene integrating into active chromatin (euchromatin) is more likely to be expressed. The genome contains regions of predominantly heterochromatin (e.g. centromeres) and predominantly euchromatin (e.g. subtelomeric regions), which show broadly consistent repressive or permissive influences on integrated transgenes.

Position effects do not necessarily affect all cells equally. In *Drosophila*, variable position effects occur naturally as a phenomenon known as position effect variegation, often as a result of chromosome rearrangements. Genes adjacent to the rearrangement may be inactivated in some cells as heterochromatin spreads across the breakpoint, but remain active in others due to the variable extent of spreading in different cells. Similar phenomena have been documented in transgenic mice resulting from the variable spreading of heterochromatin into an integrated transgene. The outcome is similar to delayed integration (Section 8.2), i.e. the transgene is expressed in only a proportion of cells in the developing animal, although in this case it is not the position of the transgene per se that is causing variable activity, but the position of the transgene relative to heterochromatin domains in different cell clones. An example of the above was observed in transgenic mice expressing β-lactoglobulin under the control of a mammary gland promoter. Discrete patches of transgene-expressing cells were observed in the transgenic mammary gland, surrounded by rings of non-expressing cells. It was proposed that patchy or variegated silencing might reflect the proximity of the transgene array to centromeric heterochromatin or, in some cases, the formation of heterochromatin de novo at highly repetitive transgene loci. (See references (7–11).)

8.5 Integration effects

Position effects are caused by the influence of the local molecular environment on transgene activity, but the transgene itself may also influence the surrounding genomic DNA resulting in so-called *integration effects*. The most obvious form of integration effect is *insertional mutagenesis*, where the transgene interrupts and hence inactivates an endogenous gene and generates a mutant phenotype. Many anecdotal accounts of accidental gene inactivation by transgene or retroviral insertion have been documented, and the process can be exploited to generate insertional mutants deliberately as described in Chapter 9. Integration effects often result from simple insertional disruption, but may also arise through the deletion of variable amounts of cellular DNA surrounding the integration site. This basis of this phenomenon is unknown although it is likely to be caused by aberrant DNA repair at the integration site. Up to 100 kb of genomic DNA may disappear.

Insertional mutagenesis is a phenomenon in which an integrating transgene disrupts and therefore alters the sequence of an endogenous gene. However, there is another class of integration effects which does not physically disrupt the affected gene, but nevertheless alters its expression pattern. For example, a transgene integrating within the range of an enhancer that serves a nearby gene may 'poach' the enhancer, i.e. prevent it from activating its normal target, resulting not only in ectopic transgene activation, but also a reduction in endogenous gene activity. In contrast, integrating transgenes can also increase the expression of a nearby gene if that gene comes under the influence of promoter or enhancer sequences carried in the transgene. Retroviral vectors are renowned for this sort of activity because they generally carry a strong, rightward-facing promoter, which can activate endogenous genes adjacent to the insertion site. Where this is undesirable, e.g. where the vectors are used for gene therapy, self-inactivating variants are used which carry mutations in the right-hand LTR (see page 91). In other cases, however, this is a desirable phenomenon. For example, *Drosophila* P-elements have been developed into specialized *activation traps* which insert randomly and activate nearby genes, providing more information on their functions (Chapter 9).

Another perhaps surprising consequence of transgene integration is the alteration of genomic methylation patterns at considerable distances from the integration site. The precise basis of this phenomenon is unclear, although it is possible that the presence of the transgene activates certain defense processes in the nucleus that scan the genome for potential invasive DNA and silence it by *de novo* methylation. (See references (12,13).)

8.6 Transgene copy number and dosage effects

The number of transgene copies integrating at a transgenic locus is highly variable and difficult to control. In some transfection methods, including electroporation and chemical methods using polyplexes, a degree of control over transgene copy number can be exercised by varying the amount of DNA used in the procedure and the precise transfection parameters. Logically, the amount of recombinant protein produced by a given transgenic locus should be proportional to the transgene copy number, but for conventional transgenes at least this appears not to be the case. Hundreds of copies of the transgene may integrate in a calcium phosphate transfection experiment whereas electroporation may generate 1–10 copies, but both methods appear to be equally efficient for the production of recombinant proteins. The lack of copy number dependence may reflect several factors:

- Transgene integrity – the large number of integrated transgenes in calcium phosphate experiments may not all be intact; indeed many are likely to be truncated and rearranged, or to bear point mutations.
- Saturation – the maximum expression level will be capped by the availability of transcription factors, copies of RNA polymerase, translational initiation factors, etc.
- Position effects – 100 copies of a transgene integrating into heterochromatin will probably be less active than a single transgene copy in a transcriptionally open chromatin domain.

- Transgene organization – transgenes in high-copy-number loci are far more likely to fall foul to repeat-induced silencing (see Sections 8.7 and 8.8).

Experiments in transgenic mice have shown that position effects are the most important factor that prevents copy number-dependence in transgene expression. If position effects are abolished through the use of a large genomic transgene or a mini-transgene flanked by insulation elements (see Section 8.10) then copy number-dependence is often partially or fully restored. Thus, dosage effects are often observed in YAC-transgenic mice. For example, a range of distinct eye phenotypes was observed in mice transgenic for a *Pax*6 YAC construct, reflecting the copy-dependent dosage of this transcription factor in the developing eye. (See reference (14).)

8.7 Transgene organization

Multicopy transgenic loci are often organized in a complex manner. The structure of such loci is often described by the term *pseudotandem array*, which means that transgenes are arranged roughly as head-to-tail and/or inverted repeats, but that not all the transgenes are necessarily intact and that individual transgenes may be interspersed with other sequences which may include one or more of the following:

- Vector DNA – this is present if whole plasmids are used for transfection and represents the backbone elements and bacterial marker genes outside the expression construct.
- Carrier DNA – this is present if carrier DNA (e.g. sheared salmon sperm DNA) has been used to bulk up the DNA component of the transfection mix.
- Local genomic DNA – this is host DNA which is obviously derived from the genomic region surrounding the insertion site itself.
- Global genomic DNA – this is host DNA which is obviously not from the region surrounding the insertion site.
- Filler DNA – this is DNA whose origin is unclear, and is probably synthesized by DNA repair enzymes during the integration process; in many cases, the filler sequences are short and are found at the junctions between plasmid sequences (plasmid–plasmid junctions) and at the junctions between plasmid and host sequences (plasmid–genomic junctions).

In cotransfection experiments (i.e. where two or more discrete transgenes are introduced into the cell on separate vectors) both transgenes tend to integrate at the same site, whether or not the two genes were originally linked on the same vector. This is further evidence that transgene arrays form episomally prior to the integration event.

The consequences of transgene organization are difficult to predict. Adjacent transgenes arranged head-to-tail (direct repeats) may influence each other's expression through promoter occlusion, if transcription reads through one transgene copy into the downstream sequence. Although polyadenylation sites should be present to prevent this, the fragmentation of vector DNA many cause two transgenes to be juxtaposed without an intervening polyadenylation site, and may facilitate the expression of fusion products caused by the unexpected joining of multiple fragments.

Adjacent transgenes arranged as inverted repeats, particularly those facing inwards (tail-to-tail) are potentially troublesome because read-through can generate hairpin RNAs that suppress transgene expression, induce the interferon response (in mammalian cells) and activate the RNA interference silencing pathway (this topic is discussed in more detail in Chapter 9).

Vector DNA can pose its own problems because certain bacterial sequences in transgenic animals have been shown to exert specific negative transcriptional effects on adjacent genes. These problems can be addressed by removing the unnecessary vector backbone and introducing only the linear transgene cassette, but even necessary sequences such as the *neo* marker can exert undesirable negative effects on adjacent promoters. (See reference (15).)

8.8 Transgene silencing

The loss of transgene expression in transformed cells and transgenic animals is generally termed *transgene silencing*. We have already described *transgene inactivation* brought about by mutation or rearrangement of the transgene during transfection, but this is a genetic process, i.e. a process in which the sequence of the transgene is altered, and is not therefore considered a true silencing mechanism. Silencing sensu stricto is an epigenetic phenomenon, i.e. the sequence of the silenced transgene is not changed. We have described in detail one form of epigenetic silencing brought about by the integration of transgenes into heterochromatin, and this can be termed *position-dependent silencing*. However, there are other forms of silencing which affect transgenes wherever they integrate, and invariably result in de novo methylation and the abolition of transgene expression.

In vertebrate genomes, approximately 10% of cytosine residues in the dinucleotide motif 5'-CG-3' are methylated. DNA methylation is often correlated with gene silencing, and transgenes methylated *in vitro* prior to gene transfer also tend to be silent, in contrast to non-methylated transgenes of identical sequence. It has been proposed that DNA methylation may be widely used as a form of gene regulation in mammals. However, except for the specialized cases of parental genomic imprinting and X-chromosome inactivation, it is thought more likely that DNA methylation is an effecter of some other primary silencing mechanism, such as altered chromatin structure. The analysis of methylation sites in mammalian DNA suggests that the primary role of DNA methylation may be to defend against invasive DNA, such as that of viruses and transposable elements. Most naturally occurring methylated cytosine residues in human DNA have been found to lie within *Alu* and LINE elements, two abundant classes of dispersed repetitive DNA corresponding to transposable elements, and it may therefore act to repress these genomic parasites and prevent them from peppering the genome with copies of themselves.

Transgenes integrating into genomic DNA sometimes undergo de novo genomic methylation and silencing even if they lie in a permissive region of the genome and have a simple structure (e.g. a single copy integration event). Many different transgenes have been affected in this manner so no specific sequence appears to be involved. This is thought to reflect the activation of a genome-scanning defense mechanism which is alerted by DNA sequences that have the wrong sequence context. Mammalian DNA has regions which are generally more GC-rich than the average, and regions that are generally more AT-rich than the average. A transgene with high GC-content integrating in an AT-rich region of the genome may well give itself away due to its unusual sequence context. This phenomenon is therefore termed *context-dependent silencing*. The *de novo* methylation of such integrated transgenes and their sequestration into heterochromatin may in some cases be followed by the spreading of inactivation into the surrounding genomic DNA, representing a further class of integration effect.

Another form of transgene silencing can be triggered by the presence of multiple copies of the same sequence, either tandemly arranged or dispersed in the genome. Such *homology-dependent silencing* occurs at two levels. At the transcriptional level, silencing often occurs if there is homology in the promoter region, and is thought to result from ectopic DNA pairing, which initiates *de novo* methylation. Cre recombinase (Chapter 6) has been instrumental in demonstrating that this phenomenon of *repeat-induced gene silencing (RIGS)*, which has been reported in *Drosophila* and mammals as well as many other organisms (*Figure 8.2*). Silencing can also occur at the post-transcriptional level, requiring homology in the coding region, and may involve the production of aberrant RNA molecules. Alternatively, a threshold system has been proposed, in which a certain mRNA dosage stimulates RNA degradation. Both these processes may ultimately reflect the production of small amounts of double-stranded RNA, e.g. by aberrant transcription of an integrated transgene or by saturation of the polyadenylation apparatus at high levels of transgene expression, resulting in the formation of a hairpin which self-primes the synthesis of a second RNA strand using mRNA as the template.

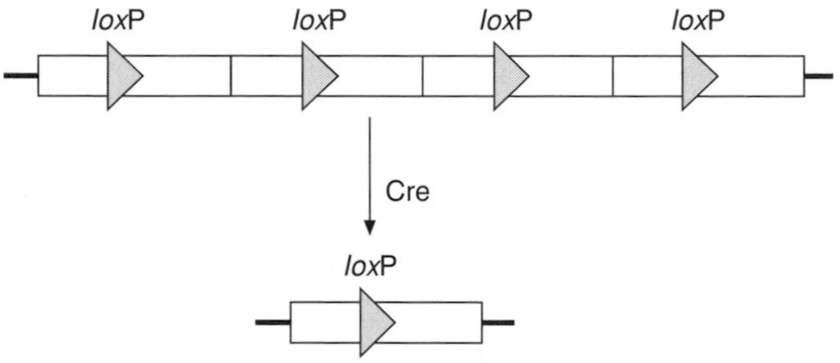

Figure 8.2

An experiment to demonstrate homology-dependent gene silencing in mammals (also called repeat-induced gene silencing (RIGS)). A transgene construct containing the human β-globin cDNA was modified to contain a single *loxP* site, which is recognized by Cre recombinase. Transgenic mice were generated carrying multiple copies of the transgene, and in these animals the locus was highly methylated and β-globin expression was low. However, when Cre recombinase was expressed, recombination between the *loxP* sites resulted in the excision of all copies of the transgene except one. Reduction in the transgene copy number resulted in increased expression, accompanied by reduced methylation at the transgenic locus.

One consequence of post-transcriptional gene silencing is a phenomenon known as *cosuppression* in which the presence of a transgene homologous to an endogenous gene can result not only in transgene silencing but also silencing of the endogenous gene. Although very widely documented in plants, the first cases of cosuppression-like silencing in animals were observed in *Drosophila* when two to six copies of a cassette containing an alcohol dehydrogenase (*Adh*)-*lacZ* fusion transgene, driven by the *white* gene promoter, were shown to reduce endogenous *Adh* gene expression both in larvae and adults. Additionally, the effect could be modified by mutations in genes such as *Polycomb*, which encode chromatin proteins involved in gene repression.

A whole industry has now developed based around the deliberate use of such phenomena to silence endogenous genes and the various different strategies for achieving this are considered in Chapter 9. (See references (16–24).)

8.9 Other problems with transgene expression

The loss of transgene expression in mammals may occur spontaneously after one or more generations and may be restricted to particular lines, suggesting that as with position effects a structural phenomenon is not to blame. Research into such unanticipated failures in transgene expression has invariably identified genetic background as the disruptive influence, as well as the parental route of transgene transmission. Transgenes, like endogenous genes, may be subject to parental imprinting effects in mammals. Therefore, the offspring of a transgenic male and wild-type female may show a different phenotype to that of a wild-type male and transgenic female, through sex-specific transgene methylation during gametogenesis. This principally affects transgenes that happen to have integrated into an imprinted region of the genome and can therefore be regarded and an additional, although rather specialized, type of position effect.

Particular mouse strains are associated with different forms of transgene behavior. BALB/c and C57BL/6 backgrounds are thought to promote *de novo* methylation, while inbred lines such as DBA/2 are thought to promote demethylation. Hence, crossing a founder transgenic mouse with a BALB/c wild-type mouse may cause transgene silencing in the mixed genetic background. Conversely, crossing to a DBA/2 mouse often results in continued transgene activity, and can also erase transgene parental imprinting effects. (See references (25–27).)

8.10 Overcoming position effects

Among the epigenetic phenomena discussed in this chapter, position effects are the simplest to overcome because the genome has its own mechanisms to avoid clashes between genes in different chromatin environments, and these mechanisms can be harnessed to protect integrated transgenes. Four different strategies have been devised to counter position effects:

- An increase in transgene copy number.
- Integration of a naked transgene at a site with beneficial position effects.
- Protection of the transgene from dominant position effects using boundary elements.

- Ensuring the transgene is equipped with its own dominant elements to establish an open chromatin domain.

The first strategy is really only relevant in cell lines which can be maintained under amplifiable selection, and where the aim is to maximize protein expression. Essentially, the approach is to increase the transgene copy number to such a massive level that, even if all the copies are minimally active, the transcriptional and protein synthesis machinery of the cell is still saturated.

The other three strategies are used in transgenic animals because they are more subtle and allow the intended expression qualities of the transgene to be preserved. The identification of a permissive locus, which allows the robust expression of even a naked transgene is one strategy, and such sites must be identified empirically. Once identified, however, *loxP* sequences can be introduced by gene targeting and transgenes can be introduced specifically at that locus by site-specific integration, or gene targeting by homologous recombination can be used to place the transgene correctly (Chapter 6). An example of such a permissive locus is the untranslated region of the *COA1A* gene, which was used for α1-antitrypsin expression in transgenic gene-targeted sheep.

The third strategy is to protect transgenes from position effects in the same way that the cell is thought to isolate active and repressed chromatin domains, by using *cis*-dominant boundary elements. It has been observed that genes *in situ* reside within large chromatin domains, which extend many kilobases either side of the transcription unit. Active chromatin domains are biochemically defined by the extent of nuclease sensitivity, and transcription per se is not required for a domain to remain in the open configuration. DNA sequences found at the domain borders often function as insulator elements, and can be used to bracket a transgene and isolate it within its own chromatin domain. Such elements have been identified, e.g. in the chicken lysozyme A gene locus and have been shown to insulate transgenes from the activity of local enhancers and to protect transgenes from position effects in cell lines and transgenic animals. Such elements are thought to act by attaching to the nuclear matrix and isolating the transgene as a topologically defined chromatin loop. In animals, it has also been observed that when such boundary elements protect transgenes from position effects, transgene expression often becomes copy number-dependent.

Instead of using flanking boundary elements, transgenes can be protected using a single copy of a dominantly acting transcriptional control element whose normal function is to establish an open chromatin domain. The most widely studied of these elements is the *locus control region* (LCR) from the β-globin locus. When β-globin transgenes are introduced into transgenic mice, robust and copy number-dependent expression is observed in lines where the transgene construct also contains a LCR, while expression is variable and often subject to silencing if the LCR is not included.

Finally in this third category, position effects can often be avoided by the use of large genomic transgenes contained in BAC or YAC vectors rather than minigenes in plasmid or retroviral vectors. Such constructs again appear to block the effects of surrounding chromatin and exhibit copy number-dependent expression. It is thought that large constructs probably contain dominant *cis*-acting elements such as insulator sequences and LCRs which provide the authentic context for transgene expression. Such elements may also reside within the introns of many genes, which would explain why intron-containing genomic expression constructs are much more robust in transgenic mice than minigenes or cDNA sequences, even if all the appropriate flanking regulatory elements are provided.

The fourth strategy for overcoming position effects is termed *transgene rescue*. An analysis of the literature on transgenic animals reveals occasional cDNA transgenes, which never appear to be subject to position effects. One example is the β-lactoglobulin (*BLG*) gene, which in most experiments shows strong and robust expression regardless of the integration site. It is presumed that, like the large transgene constructs mentioned above, this cDNA transgene carries within it some functional sequence that establishes an open chromatin domain and blocks the spreading of surrounding hetero-chromatin. The idea behind transgene rescue is that a primary experimental transgene, which is not able to withstand position effects, is introduced along with the *BLG* gene, and exploits the permissive environ-ment established by the latter. (See references (28–34).)

References

1. Iyengar A, Muller F, Maclean N (1996) Regulation and expression of transgenes in fish – A review. *Transgenic Res* **5**: 147–166.
2. Kothary RK, Allen ND, Barton SC, Norris ML, Surani MAH (1992) Factors affect-ing cellular mosaicism in the expression of a lacz transgene in 2-cell stage mouse embryos. *Biochem Cell Biol* **70**: 1097–1104.
3. Miller CK, Temin HM (1983) High-efficiency ligation and recombination of DNA fragments by vertebrate cells. *Science* **200**: 606–609.
4. Bishop JO, Smith P (1989) Mechanism of chromosomal integration of microin-jected DNA. *Mol Biol Med* **6**: 283–298.
5. Stuart GW, Vielkind JR, McMurray JV, Westerfield M (1990) Stable lines of trans-genic zebrafish exhibit reproducible patterns of transgene expression. *Development* **109**: 577–584.
6. Endean D, Smithies O (1989) Replication of plasmid DNA in fertilized xenopus eggs is sensitive to both the topology and size of the infected template. *Chromosoma* **97**: 307–314.
7. Al-Shawi R, Kinnaird J, Burke J, Bishop JO (1990) Expression of a foreign gene in a line of transgenic mice is modulated by a chromosomal position effect. *Mol Cell Biol* **10**: 1192–1198.
8. Elgin SCR (ed.) (1995) Chromatin Structure and Gene Expression. IRL Press, Oxford, UK.
9. Tsukiyama T, Wu C (1997) Chromatin remodelling and transcription. *Curr Opin Genet Dev* **7**: 242–248.
10. Bestor TH (1998) Gene silencing – Methylation meets acetylation. *Nature* **393**: 311–312.
11. Dobie K, Mehtali M, McClenaghan M, Lathe R (1997) Variegated gene expres-sion in mice. *Trends Genet* **13**: 127–130.
12. Doerfler W, Schubbert R, Heller H, Kammer C, Hilger-Eversheim K, Knoblauch M, Remus R (1997) Integration of foreign DNA and its consequences in mammalian systems. *Trends Biotechnol* **15**: 297–301.
13. Rijkers T, Peetz A, Ruther U (1994) Insertional mutagenesis in transgenic mice. *Transgenic Res* **3**: 203–215.

14. Schedl A, Ross A, Lee M, Engelkamp D, Rashbass P, Vanheyningen V, Hastie HD (1996) Influence of PAX6 gene dosage on development: Over expression causes severe eye abnormalities. *Cell* **86**: 71–82.

15. Artelt P, Grannemann R, Stocking C, Friel J, Bartsch J, Hauser H (1991) The prokaryotic neomycin-resistance encoding gene acts as a transcriptional silencer in eukaryotic cells. *Gene* **99**: 249–254.

16. Jahner D, Stuhlmann H, Stewart CL, Harbers K, Lohler J, Simon I, Jaenisch R (1982) De novo methylation and expression of retroviral genomes during mouse embryogenesis. *Nature* **298**: 623–628.

17. Jahner D, Jaenisch R (1985) Retrovirus-induced de novo methylation of flanking host sequences correlates with gene inactivity. *Nature* **315**: 594–597.

18. Kumpatla SP, Chandrasekharah MB, Iyer LM, Li G, Hall TC (1998) Genome intruder scanning and modulation systems and transgene silencing. *Trends Plant Sci* **3**: 97–104.

19. Clark AJ, Harold G, Yull FE (1997) Mammalian cDNA and prokaryotic reporter sequences silence adjacent transgenes in transgenic mice. *Nucleic Acids Res* **25**: 1009–1014.

20. Gallie DR (1998) Controlling gene expression in transgenics. *Curr Opin Plant Biol* **1**: 166–172.

21. Grant SR (1999) Dissecting the mechanisms of post-transcriptional gene silencing: divide and conquer. *Cell* **96**: 303–306.

22. Plasterk RHA, Ketting RF (2000) The silence of the genes. *Curr Opin Genet Dev* **10**: 562–567.

23. Hammond SM, Caudy AA, Hannon GJ (2001) Post-transcriptional gene silencing by double-stranded RNA. *Nature Rev Genet* **2**: 110–119.

24. Garrick D, Fiering S, Martin DI, Whitelaw E (1998) Repeat-induced gene silencing in mammals. *Nature Genet* **18**: 56–59.

25. Koetsier PA, Mangel L, Schmitz B, Doerfler W (1996) Stability of transgene methylation patterns in mice – Position effects, strain specificity and cellular mosaicism. *Transgenic Res* **5**: 235–244.

26. Reik W, Howlett SK, Surani MA (1990) Imprinting by DNA methylation – From transgenes to endogenous gene-sequences. *Development (suppl.)*: 99–106.

27. Sasaki H, Hamada T, Ueda T, Seki R, Higashinakagawa T, Sakaki Y (1991) Inherited type of allelic methylation variations in a mouse chromosome region where an integrated transgene shows methylation imprinting. *Development* **111**: 573–581.

28. Sauer B (1995) Site-specific recombination: developments and applications. *Curr Opin Biotechnol* **5**: 521–527.

29. Stacey AJ, Schnieke A, McWhir J, Cooper J, Colman A, Melton DW (1994) Use of double-replacement gene targeting to replace the murine alpha-lactalbumin gene with its human counterpart in embryonic stem cells and mice. *Mol Cell Biol* **4**: 1009–1016.

30. Geyer PK (1997) The role of insulator elements in defining domains of gene expression. *Curr Opin Genet Dev* **7**: 242–248.

31. Stief A, Winter DM, Stratling WH, Sippel AE (1989) A nuclear-DNA attachment element mediates elevated and position-independent gene activity. *Nature* **341**: 343–345.

32. McKnight RA, Shamay A, Sankaran L, Wall RJ, Hennighausen L (1992) Matrix-attachment regions can impart position-independent regulation of a tissue-specific gene in transgenic mice. *Proc Natl Acad Sci USA* **89**: 6943–6947.

33. Ellis J, Pasceri P, Tan-Un KC, Wu X, Harper A, Fraser P, Grosveld F (1997) Evaluation of beta-goblin gene therapy constructs in single copy transgenic mice. *Nucl Acids Res* **25**: 1296–1302.

34. Clark AJ, Archibald AL, McClenaghan M, Simons JP, Wallace R, Whitelaw CBA (1993) Enhancing the efficiency of transgene expression. *Phil Trans R Soc Lond B* **339**: 225–232.

Additional strategies for gene inactivation

9

9.1 Introduction

Eliminating the expression of endogenous genes can be achieved by gene targeting or a combination of gene targeting and site-specific recombination as discussed in Chapter 6. However, routine gene targeting is only possible in accessible cells that readily undergo homologous recombination, and at the present time this is largely restricted to murine ES cells, *Drosophila* embryos and a few obscure somatic cell lines. Any cell could probably be used for gene targeting if the investigator were willing to expend the effort, time and resources needed to identify ultra-rare recombinants in large background populations. However, for today's high-throughput experiments where the aim is to annotate the thousands of anonymous genes discovered in the genome projects, much simpler and more scalable methods are required and such methods must be applicable in all animals. Those methods are the subject of this chapter.

The creation of a functional gene product is a long and complex process, which can be divided into three major stages: the DNA stage (transcription), the RNA stage (RNA processing) and the protein stage (protein synthesis, modification and localization). Strategies for gene inactivation are available that disrupt each of these stages, and we discuss them in turn.

At the DNA stage, a widely applicable strategy for gene inactivation is insertional mutagenesis. Indeed, this approach is possible in any animal or animal cell which is amenable to gene transfer. Insertional mutagenesis is the use of DNA cassettes to generate mutants by random insertion into existing genes. The cassettes can be used to generate thousands of mutants, and then the mutated genes can be recovered using simple cloning strategies. An extension of insertional mutagenesis is the use of entrapment vectors, which also insert randomly into the genome and generate mutant phenotypes, but have the added ability to report the expression profile and other useful information about each interrupted gene.

Strategies that interfere with gene expression at the RNA and protein levels do not alter the DNA sequence of the target gene, but instead either destroy the RNA or inhibit the activity of the protein. Many different techniques can be used to inhibit gene expression at the RNA level, some of which have a transient impact and some of which result in permanent gene inactivation. Transient methods include transfection with antisense DNA or RNA oligonucleotides, while permanent methods include the production of transgenic animals expressing antisense RNA or ribozymes. The most significant recent development in RNA-based gene inhibition methods is RNA interference (RNAi), a potent silencing method, which exploits a ubiquitous cellular defense mechanism that protects cells from viruses.

RNAi has risen from relative obscurity to become one of the hottest topics in molecular biology, and is now being used for the systematic functional analysis of whole genomes. In the latest series of RNAi studies in the nematode worm *Caenorhabditis elegans*, more than 18 000 genes have been inactivated in a single, heroic experiment.

There are also several different approaches that can be used to interfere directly with the activity of a protein. In the last part of the chapter, we discuss the use of intracellular antibodies (intrabodies), protein-disrupting oligonucleotides (aptamers) and mutant versions of proteins, which sequester the functional protein into an inactive complex (dominant negatives).

9.2 Gene inactivation at the DNA level – insertional mutagenesis

Although gene targeting and site-specific recombination can be used to disrupt specific target genes with great accuracy (at least in mice and *Drosophila*) such methods are not useful for high-throughput studies because a different targeting vector is required for each different gene. Systematic gene targeting has been achieved in the yeast *Saccharomyces cereivisae*, which has about 6000 genes, but scaling this up to *Drosophila* (18 000 genes) and the mouse (30 000 genes) is currently unthinkable, particularly because of the inefficiency of gene targeting in these species compared to yeast[1].

For large-scale mutagenesis, also known as saturation mutagenesis, investigators have long relied on radiation and chemical mutagens. These produce mostly point mutations, so the task of isolating the mutated genes can be long-winded, particularly in animals that are not amenable to genetic analysis. A breakthrough came with the development of techniques in which DNA was used as an *insertional mutagen*. This approach evolved from the observation of natural phenomena such as hybrid dysgenesis in *Drosophila* (page 122), which is caused by the random insertion of transposable elements into active genes and their consequent disruption and inactivation. Similarly, about 5% of naturally occurring recessive mutations in the mouse are caused by retroviral insertion. By harnessing such elements, it became possible to saturate the genomes of many animals with insertions and isolate mutations affecting many genes.

The use of endogenous transposable elements has several disadvantages. Some animals do not contain active transposons (e.g. in most fish, the transposons that are present in the genome appear to be internally deleted and therefore inactive), while in other animals there may be thousands of copies of a given element, making it difficult in many cases to identify which gene has been interrupted. It is here that gene transfer can help, because exogenous DNA can also be used as an insertional mutagen, providing a unique insertional sequence that can be mapped and cloned. Two strategies have been developed:

---

1 In contrast to the situation in animal cells, gene targeting in yeast is more efficient than random integration and is therefore the most likely outcome of a gene transfer experiment if the exogenous DNA is homologous to a yeast gene.

- Mutagenesis with non-mobile exogenous DNA. In this approach, trans-genes introduced into the recipient genome insert into exogenous genes and remain immobilized at that position. Interestingly, insertional mutants have often been identified as by-products of other gene transfer experiments, when the transgene happens to integrate into another gene, disrupting it and generating a mutant phenotype.
- Mutagenesis with heterologous or artificial transposons. P-elements in *Drosophila* have been introduced into M-strains, which lack such elements, and used for insertional mutagenesis. The *Sleeping Beauty* transposon system is a synthetic system developed for fish (page 121).

The advantage of foreign or synthetic transposable elements or transgenes (compared to endogenous elements) is that the mutated gene becomes tagged with a unique DNA sequence, which can then be exploited to clone the interrupted gene. A straightforward but rather laborious method is to create a genomic library from the mutated cells or animals and isolate the clone containing the insertion by hybridization or PCR. Although the insert is the handle used to isolate such clones, some flanking DNA from the interrupted locus is inevitably linked to the insert, and this can be sequenced and used to identify the interrupted gene. Rather than creating a complete genomic library, a shortcut is to use a technique known as *plasmid rescue* in which the insertion element contains the origin of replication and antibiotic selection marker from a bacterial plasmid. Genomic DNA isolated from any interesting mutant can then be digested with an appropriate restriction enzyme, diluted and self-ligated to form circular genomic fragments, and used to transform bacteria. Only circles containing the plasmid maintenance sequences will replicate in bacteria under selection, while the remaining genomic fragments will be lost. The rescued plasmid will also contain flanking sequences which can be used to identify the interrupted gene (*Figure 9.1*).

An even greater simplification is the inverse PCR technique and its relatives, all of which exploit the principle of *outward-facing* primers to amplify the flanking sequences directly, therefore allowing all analysis steps to be carried out *in vitro*. The basic I-PCR technique is shown in *Figure 9.2*. Genomic DNA from the mutant cell line or animal is digested with a restriction enzyme for which there is no site in the insertional tag. The genomic fragments are then self-ligated to form circles, and one of these circles will contain the insertion element and some flanking DNA. PCR primers annealing at the edges of the insertion element but facing outwards can then be used to amplify the flanking genomic sequences, which as above can be used to identify or characterize the interrupted gene. (See references (1–4).)

9.3 Entrapment vectors

A more sophisticated way to use insertional vectors is to provide them with modules that yield functional information about the interrupted gene. These modified insertional elements are called *entrapment vectors* because they are activated only when they integrate at the appropriate position within or adjacent to an endogenous gene. Essentially, they exploit the *position effects* phenomenon which is described in Chapter 8 – they carry weak

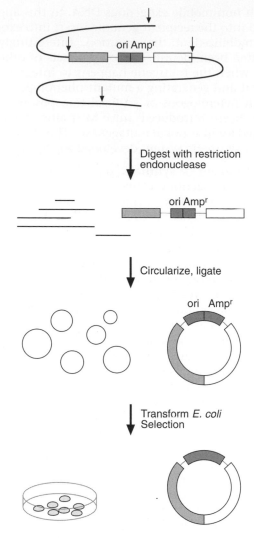

Figure 9.1

The principle of plasmid rescue, a technique for isolating genomic DNA sequences flanking a transposon or transgene integration site.

regulatory elements which respond to genomic regulatory elements in the local environment, and they contain a reporter gene which divulges the pattern of expression driven by those genomic elements, and in some cases further information about the gene product.

Enhancer traps

The first entrapment vectors were *enhancer traps* based on *Drosophila* P-elements. The original enhancer traps were equipped with the reporter

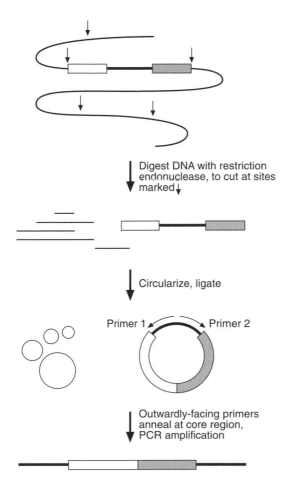

Figure 9.2

Inverse PCR. The insert is indicated by the thick line. Restriction sites are marked with arrows, and the left and right flanking regions in the genomic DNA are represented by closed and open boxes. Primers are designed to hybridize with the insert and are extended in the directions shown. PCR amplification generates a linear fragment containing left and right flanking sequences.

gene *lacZ* transcribed from a minimal promoter comprising a TATA box and initiator, so that the basal transcriptional activity of the construct was very low. When such elements were introduced into the *Drosophila* germline using the methods described in Chapter 5, they integrated at many different genomic sites. In some cases, an enhancer trap might happen to integrate within the range of an endogenous enhancer, and this would stimulate the activity of the promoter. In such flies, the expression profile of *lacZ*, revealed as a pattern of blue cells when larvae or adult flies were incubated in X-gal, would report the expression profile specified by the enhancer (*Figure 9.3*).

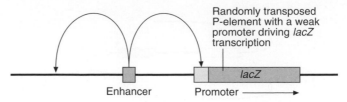

Figure 9.3

Enhancer trapping. A reporter construct consists of *lacZ* linked to a weak promoter, which requires an enhancer for significant transcriptional activity. P-element-mediated transposition is used to insert the construct at random sites in the fly genome. When the promoter is inserted within the active range of an enhancer, expression of *lacZ* can be detected.

Enhancers are usually located quite some distance from the genes they regulate, so enhancer traps have rarely been used for the isolation of genes. However, they can be exploited to drive other transgenes in cell-specific patterns. This is achieved using a two-step system in which the enhancer trap construct encodes not a reporter gene, but a yeast transcription factor such as GAL4 that can be used to activate any gene carrying the appropriate *cis*-acting elements in its promoter. As an example, consider a situation where it is desirable to express a toxin gene in a specific population of brain cells and therefore ablate those cells and investigate their role in fly development. An enhancer trap screen is set up by transforming flies with a P-element in which the weak promoter drives the expression of GAL4. Although the expression pattern cannot be observed directly, it can be seen if the enhancer trap flies are crossed to a strain containing the *lacZ* gene regulated by a promoter responsive to GAL4. In this second generation of flies, the observation of individuals with the correct pattern of blue-staining cells would allow the parental strain, in which the transcription factor is expressed in those cells, to be identified. This parental line can now be crossed to another line, in which a toxin gene such as ricin is controlled by a GAL4-responsive promoter. In this way, banks of fly lines with the transcription factor expressed in different spatial and temporal patterns can be established, and used to drive any desired transgene in any desired pattern (*Figure 9.4*). Since the development of the original *Drosophila* system, enhancer traps based on similar principles have also been used in mice and fish.

Gene traps

The gene trap is an entrapment vector that is only activated when it inserts *within* a gene. Like the enhancer trap, the gene trap consists of a reporter gene controlled by minimal regulatory elements, which can vary in their exact nature as discussed in more detail below. Because genes make up only about 3% of a mammalian genome, the gene trap approach makes insertional mutagenesis much more efficient, since insertions that fail to activate the gene trap can be set aside and the investigator can focus on an enriched collection of insertions that definitely represent interrupted genes. Those

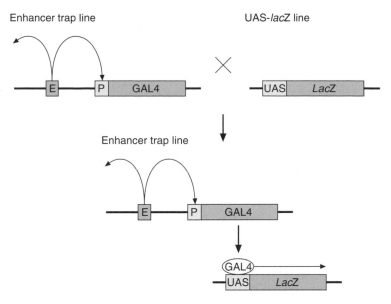

Enhancer trap line   UAS-*lacZ* line

Figure 9.4

Second-generation enhancer trap, in which the yeast gene for the GAL4 tran-scription factor is activated by the enhancer. This can be crossed to a responder line in which *lacZ* is driven by a GAL4-dependent promoter.

with interesting expression patterns can be selected for further analysis, and the insert can be used to clone the flanking sequences as discussed earlier and identify the interrupted gene.

Several types of gene trap vector have been developed, which rely on different components of the interrupted gene to activate the reporter:

- Promoter trap (*Figure 9.5*). This construct comprises a naked reporter gene with or without an AUG codon, and can only be activated if it inserts within an exon of an expressed gene, in the correct orientation and in the correct reading frame relative to the surrounding gene. It is expressed as a fusion transcript. If an AUG is present, independent trans-lation may take place, but if no AUG is present it can only be expressed as a fusion protein.
- Splice trap (*Figure 9.5*). This comprises a reporter gene downstream of a splice acceptor site. If it inserts within an intronless gene, the splice site is ignored and the construct is activated in the same way as a promoter trap and under the same constraints. If it inserts within an intron or exon of a split gene, then the reporter gene may be spliced into the endogenous gene product and may be expressed if it is inserted in the correct orientation and reading frame.
- IRES trap. These are splice traps in which the reporter gene is driven by an IRES element, which means that there is no dependence on in-frame insertion.
- Poly(A) trap (*Figure 9.6*). These are very sophisticated IRES splice traps, which also contain a selectable marker driven by a constitutive promoter,

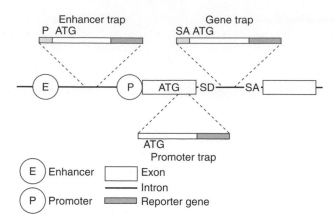

Figure 9.5

Comparison of the structure and typical integration positions of three different types of entrapment vector – the enhancer trap, a typical gene (splice) trap and a promoter-dependent gene trap (promoter trap). SA = splice acceptor, SD = splice doner and ATG = translational start site.

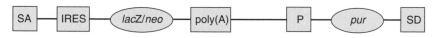

Figure 9.6

The highly versatile gene trap vector designed by Zambrowicz and colleagues (13) for high-throughput gene trapping in the mouse. The splice acceptor site (SA) is upstream of an internal ribosome entry site (IRES) and the β-Geo reporter gene (*lacZ/neo*), which has its own poly(A) site. Therefore, expression of the visible reporter is dependent on integration into an endogenous gene, which is expressed in ES cells, but is not dependent on in-frame insertion. Selection for gene insertions is based on puromycin resistance. The *pur* selectable marker is controlled by a constitutive promoter (P), and is therefore independent of the activity of the interrupted gene, but is dependent on that gene's polyadenylation site. SD = splice donor.

but polyadenylation of the selectable marker is dependent on the interrupted gene. This means that insertion events can be identified by selection even if the interrupted gene is not expressed at the time of analysis.

- Cre trap. A binary system in which the gene trap contains a *cre* transgene activated by insertion into an endogenous gene, and a *lacZ* reporter separated from a constitutive promoter by a blocking sequence which is flanked by *loxP* sites. If the gene trap is activated, Cre recombinase is expressed, and removes the block and allows expression of *lacZ*. This means insertion events can be identified even if the interrupted gene is expressed transiently.

Gene trapping has been widely used in the functional analysis of genomes, in mice, *Drosophila* and most recently in *Xenopus laevis*. For example, several laboratories have now produced large libraries of gene-trapped ES cell lines using retroviral gene trap vectors, and have carried out systematic analyses to identify genes with particular expression patterns in undifferentiated ES cells or cells which have been stimulated to differentiate in various ways. The *Xenopus* gene trap uses green fluorescent protein as the reporter system allowing expression profiles to be analyzed in living tadpoles in real time.

Activation traps

The activation trap is a rather different form of entrapment vector whose purpose is to generate gain-of-function mutations not by integrating into genes but by integrating next to them and increasing their activities. This is achieved by equipping the constructs with a strong, outward facing promoter. Activation traps are useful for two major reasons. First, it is difficult to achieve saturation with conventional insertional mutagenesis because the insertion events are not truly random. In *Drosophila* gene trap screens, for example, it has been estimated that up to 50% of genes are never tagged by insertions for this reason. However, since activation traps do not need to insert into the gene but can integrate a considerable distance away, many of these recalcitrant loci can be investigated. Second, insertions often fail to generate informative phenotypes. Reasons for this include:

- The insertion may occur in a non-essential region of the gene.
- If a locus is haplosufficient, heterozygous loss will not generate a phenotype.
- Homozygous loss of function may generate no phenotype if the gene product is not required under the experimental conditions used.
- Homozygous loss of function may generate no phenotype if the gene product is functionally redundant with another.

However, the overactivity or ectopic activity of a gene can often generate a phenotype where the loss of activity cannot, so both gene trap and activation trap strategies are currently being explored in the *Drosophila* gene trap program. (See references (5–18).)

9.4 RNA-based strategies for transgene inactivation

We turn now to methods which are used to inactivate genes by interfering with gene expression at the RNA level. Such methods involve the introduction of inhibitory DNA or RNA molecules into the cell, or the insertion of transgenes which produce inhibitory RNA molecules as a product (*Table 9.1*). These molecules interact directly either with the mRNA of the target gene or, more rarely, with the gene itself. However, the sequence of the target gene is never modified, so it is incorrect to use terms such as mutation and mutagenesis to describe the techniques, and terms such as interference, inactivation or inhibition are preferred. It is also incorrect to describe the effects of such procedures as changes in phenotype – strictly speaking, the resulting loss-of-function effects are termed *functional knockouts* or *phenocopies*.

Table 9.1 Nucleic acid interfering entities used to inhibit transgene expression

Interfering entity	Description and inhibitory mechanism
Antisense RNA	RNA with a complementary sequence to mRNA. Blocks translation and promotes degradation.
Antisense transgene	A transgene which is inverted with respect to its promoter, and therefore produces antisense RNA instead of mRNA.
Antisense oligonucleotide	A short, single stranded DNA or RNA molecule (typically less than 25 nucleotides) which is complementary to a given mRNA. Can act in the same manner as antisense RNA or may form a triplex in the gene.
Ribozyme	An antisense RNA molecule with a catalytic center that can cleave the target mRNA molecule.
Deoxyribozyme	The DNA equivalent of a ribozyme, functioning in the same manner as above.
Maxizyme	A ribozyme whose activity can be controlled by a small allosteric modulator – an inducible ribozyme. Regulated RNA lassoes are based on a similar principle.
Sense RNA	A synthetic RNA molecule with the same sequence as mRNA. May promote the formation of dsRNA molecules by snap-back synthesis thus inducing RNA interference.
dsRNA	Double stranded RNA, the trigger for RNA interference.
siRNA	Small interfering RNA – short dsRNA molecules produced by the cleavage of longer dsRNA by the enzyme Dicer, which provide the specificity for RNA interference.
SIRPLEX	A chemically synthesized siRNA (short interfering RNA duplex).
Hairpin RNA	A double stranded RNA molecule covalently sealed at one end, expressed from an inverted-repeat transgene with the specific intention of inducing RNA interference.
Aptamers	DNA or RNA molecules than bind specifically to proteins and inhibit their activities.
Intramers	RNA aptamers expressed inside the cell from a small transgene.
Spiegelmers	Aptamers made from L-oligonucleotides rather than the natural R-oligonucleotides. These are resistant to cellular nucleases and therefore have a longer half-life.
Decoy RNA	An RNA molecule that inhibits the activity of an RNA-binding protein by providing a decoy binding site and sequestering it from its true substrate.

Antisense and sense RNA

Antisense RNA has the opposite sense (or complementary sequence) to mRNA, and the two molecules can form stable duplexes when present in the same cell. The formation of such a duplex may inhibit transcription, RNA processing and translation, combining to strongly inhibit the expression of the endogenous gene. Transient antisense effects are apparent if cells are transfected with synthetic antisense RNA or oligonucleotides, but permanent inhibitory effects can be produced if such molecules can be expressed within the cell by transformation with an antisense transgene. Antisense transgenes are constructed simply by inverting the coding sequence of the target gene with respect to the vector's promoter, and then introducing the construct into the cell in the conventional manner.

The first antisense experiments in transgenic animals were carried out in 1988 by Katsuki and colleagues, and involved the transformation of mice with a construct containing the complementary sequence to part of the myelin basic protein gene, *shiverer*. Many of the resulting transgenic mice were heavily depleted for the protein, and in the most extreme cases only about 20% of the normal amount of protein was produced, generating a close phenocopy of the genuine shiverer phenotype. Many other genes have been inhibited in the same manner, both in cell lines and in transgenic animals. The efficiency of the technique varies considerably and may depend on which part of the gene is targeted. Some investigators have used full-length antisense constructs while others have reported successful antisense inhibition with antisense transgenes less than 40 nucleotides in length. There have also been several reports of conditional antisense inhibition, reflecting the use of inducible promoters to control antisense expression.

While the mechanism by which antisense RNA inhibits gene expression seems logical, the ability of sense RNA (with the same sequence as mRNA) to exert the same effect is more difficult to explain. The ability of sense transgenes to inhibit the activity of homologous endogenes was first demonstrated in plants, where the process can be extremely potent, in some cases reducing both transgene and endogenous gene expression to zero. This effect has been termed *cosuppression*. It was several years after the discovery of this phenomenon in plants that a similar process was observed in animals, initially in *Drosophila* and worms, and then in mammals. The degree of silencing is never as strong as that seen in plants, so cosuppression is therefore not used as a deliberate method of gene silencing in animals.

Ribozyme constructs

Ribozymes are naturally occurring catalytic RNA molecules that cleave target RNA molecules at specific sites. Ribozyme catalytic moieties can be engineered into antisense transgenes so that the resulting antisense RNA not only binds to its mRNA substrate and inhibits translation, but also cleaves it in two, resulting in its rapid degradation. While the inhibitory effect of conventional antisense RNA is stoichiometric, ribozymes release their substrate after cleavage and are therefore capable of destroying many hundreds of target mRNA molecules, making them much more potent silencers of gene expression.

Ribozyme-based silencing constructs were first used in *Drosophila* to inhibit the *white* gene, producing flies with reduced pigmentation in the eyes. However, there have been relatively few further examples of this approach being used in transgenic animals. One interesting example in mice involved the expression of a ribozyme-based construct against glucokinase, which was expressed under the control of the insulin promoter, resulting in specific inhibition of the endogenous gene in the pancreas. The most widespread use of ribozyme constructs has been in the development of cancer therapies and strategies to tackle HIV, particularly through the use of vectors carrying several ribozymes targeting different parts of the HIV genome.

RNA interference

RNA interference (RNAi) is a potent form of gene silencing induced by the presence of double-stranded RNA (dsRNA). The phenomenon was discovered by Fire and Mehlo in 1998 while investigating the use of antisense and sense RNA for gene inhibition in the nematode worm *Caenorhabditis elegans* (31). In one experiment, they introduced both sense and antisense RNA into worms simultaneously and observed a striking and specific inhibitory effect, which was approximately tenfold more efficient than either single RNA strand alone. Further investigation showed that only a few molecules of dsRNA were necessary for RNAi in *C. elegans*, indicating that, like the action of ribozymes, the effect was catalytic rather than stoichiometric. Interference could be achieved only if the dsRNA was homologous to the exons of a target gene indicating that it was a post-transcriptional process. RNA interference occurs in all animals studied to date, and is now beginning to be used as a systematic silencing method in *Drosophila*, mice and human cells. In *C. elegans*, RNAi is now the standard procedure for gene inactivation, and it is becoming increasingly favored in other organisms due to its potency and specificity.

The mechanism of RNA interference is complex and not completely understood, but involves the degradation of the dsRNA molecule into short duplexes, about 21–25 bp in length with 2–3 nucleotide overhangs, by a dsRNA-specific endonuclease called Dicer (*Figure 9.7*). The short duplexes are known as small interfering RNAs (siRNAs). These separate into single strands and assemble an RNA-induced silencing complex, which binds to the corresponding mRNA and cleaves it into fragments, which are rapidly degraded. The process is extraordinarily effective and reduces the mRNA of most genes to undetectable levels.

In the first RNAi experiments, dsRNA was injected directly into worms. Because RNAi is a systemic phenomenon, any injection site is suitable and the effect spreads throughout the body. The germ cells are affected and the inhibitory effect persists into the next generation of worms. However, other methods of administration appear to work just as effectively. Worms can be soaked in a dsRNA solution or even fed on bacteria producing dsRNA. The introduction or expression of long dsRNA molecules is suitable for gene silencing in most animals and in mammalian embryos and embryonic cell lines. However, the results of RNA interference are masked in adult

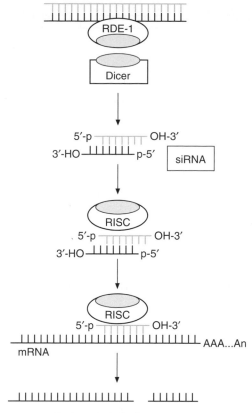

Figure 9.7

Proposed mechanism of RNA interference. A dsRNA molecule is cleaved by the enzyme Dicer into siRNA, which is double-stranded and has 3′ overhangs of two nucleotides. This is incorporated into the RNA-induced silencing complex (RISC). The siRNA serves as a guide for RISC and, upon perfect base pairing, the target mRNA is cleaved in the middle of the duplex formed with the siRNA. Modified from Voinet (2002), with permission from Elsevier Science.

mammalian cells by the interferon response, which is a general (not sequence-specific) response to dsRNA molecules over about 30 bp in length, in which protein expression is generally repressed in order to inhibit viral replication.

This problem has been circumvented by the direct administration of chemically synthesized siRNAs. Such molecules are very effective inducers of RNAi in mammalian cells, but they are also prohibitively expensive. Several strategies have therefore been developed for the *in vitro* enzymatic synthesis of siRNAs, which involve either separate transcription of sense and antisense strands followed by annealing, or the *in vitro* transcription of short hairpin RNAs (*Figure 9.8*). When such *in vitro* synthesized RNAs have

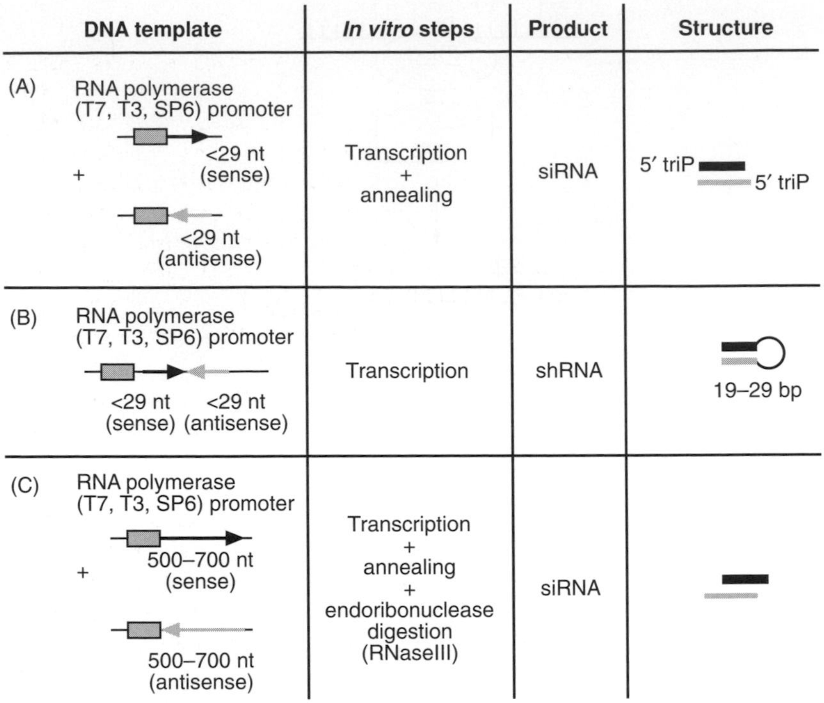

DNA template	*In vitro* steps	Product	Structure
(A) RNA polymerase (T7, T3, SP6) promoter <29 nt (sense) + <29 nt (antisense)	Transcription + annealing	siRNA	5' triP ▬▬ 5' triP
(B) RNA polymerase (T7, T3, SP6) promoter <29 nt (sense) <29 nt (antisense)	Transcription	shRNA	19–29 bp
(C) RNA polymerase (T7, T3, SP6) promoter 500–700 nt (sense) + 500–700 nt (antisense)	Transcription + annealing + endoribonuclease digestion (RNaseIII)	siRNA	▬▬

Figure 9.8

Three different approaches for the production of siRNA in vitro. (A) Sense and antisense siRNAs can be transcribed from separate promoters and annealed *in vitro*. (B) RNA polymerases T7 or T3 can be used to direct the transcription of small inverted repeats separated by a spacer region of varying lengths. The resulting RNAs form hairpins containing ~29 nt stems that match the target sequence. (C) siRNAs are produced after digestion of long dsRNAs (synthesized and annealed *in vitro*) with a purified *E. coli* RNase III.

been used in worms, they have long-lasting effects and it is thought that RNAi potency in these organisms owes much to an intrinsic amplification system, which prolongs the longevity of the effect. This amplification system appears to be absent from mammalian cells, so strategies have been developed to synthesize siRNA *in vivo* from mini-transgenes (*Figure 9.9*). In the first approach, a plasmid is constructed which contains two RNA pol III transcription units in tandem, each producing one of the siRNA strands. The separate strands are thought to assemble spontaneously into the siRNA duplex *in vivo*. The second approach is very similar, but the pol III transcription units are supplied on separate vectors. The individual RNA strands assemble in the same manner. In the third strategy, the plasmid produces a hairpin RNA which assembles into siRNA by self-pairing. Either a pol III or a pol II transcription unit can be used for this purpose. Note, however, that pol III transcription terminates at a run of thymidine residues, whereas to avoid polyadenylation of the pol II transcript, the pol II vector must be linearized at the end of the hairpin sequence prior to transfection. (See references (19–42).)

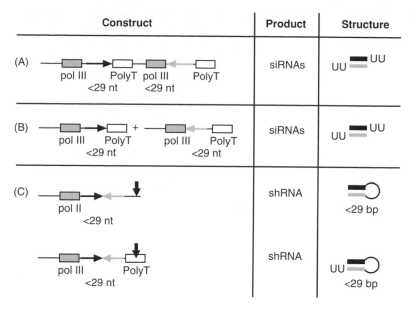

Construct	Product	Structure

Figure 9.9

Producing siRNA *in vivo*: (A) Sense and antisense strands can be transcribed *in vivo* from two independent RNA polymerase III promoters placed in tandem on the same plasmid, or (B) on two separate plasmids. (C) shRNA can be produced *in vivo*, after transcription of a sequence containing the sense and antisense strands placed in tandem downstream of the RNA polymerase pol II. In this case, the plasmid is linearized before transfection into the cells. The sequence can also be placed downstream of a RNA polymerase pol III promoter. In this case, the transcription terminates when the RNA polymerase recognizes a sequence of four consecutive thymidines.

9.5 Gene inhibition at the protein level

Oligonucleotides fold up to adopt a particular tertiary conformation in solution and can interact with proteins and in some cases inhibit their activities. The selection of suitable oligonucleotides can be carried out *in vitro* using affinity procedures, and then suitable molecules can be chemically synthesized and injected into cells. Such inhibitory molecules are known as aptamers. Similarly, antibodies are proteins that bind with great specificity and affinity to particular target antigens. Their natural function is to recognize and neutralize foreign antigens in the body and this can be exploited to inactivate specific target proteins in experimental settings. Thus, the microinjection of antibodies into cells has been widely used to block protein activity. In the case of both antibodies and aptamers, however, any inhibitory effect is short-lived.

Longer-term inhibition can be achieved by expressing antibodies or aptamers within cells from specially designed transgenes. In the case of aptamers, very similar transgene cassettes can be used as those required for siRNA expression, and the resulting molecules are known as *intramers*. However, matters become more complex for the expression of antibodies, since native serum antibodies are large tetrameric proteins comprising two heavy chains and two light chains, each with an antigen-binding domain and various constant domains that perform effecter functions. Since the effecter functions serve no useful purpose inside the cell, a useful strategy has been to clone the heavy and light chain cDNAs from specific hybridoma cells, in order to produce antibodies with the correct antigen-binding specificity, and then combine the antigen-binding regions of the heavy and light chains into a single polypeptide chain, known as a single chain fragment variable (scFv) (*Figure 9.10*). Such recombinant antibodies can be expressed from mini-transgenes within the cell, and are hence known as *intrabodies*. They bind to their target antigens with the same specificity as the parent immunoglobulin, and inhibit their activity to the same degree.

Another method for inhibiting proteins is to express a *dominant negative mutant* in the cell. A dominant negative is a non-functional mutant version of the protein, which sequesters functional versions of the protein into inactive complexes. This strategy works only for proteins that normally exist as dimers or multimers, and is particularly effective for inactivating cell surface receptors. The overexpression of a dominant negative transgene overwhelms the cell with mutant copies, mopping up all functional copies of the protein into inactive complexes. The dominant negative strategy has been used extensively to study signal transduction, development and cell–cell communication. (See references (43–48).)

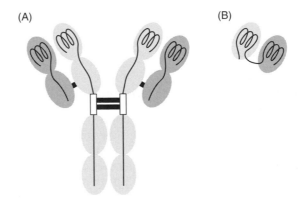

Figure 9.10

(A) The structure of a typical serum immunoglobulin, comprising two heavy chains (light gray) and two light chains (dark grey) joined by disulfide bonds (thick lines). The variable domains are shown by the curly lines. (B) In contrast, the single-chain Fv fragment is a recombinant derivative which retains one copy of each variable domain, joined by a flexible peptide linker, but no constant regions. The scFv fragment retains the antigen-binding specificity of the parent immunoglobulin but is smaller, encoded by a single transgene, does not need to form disulfide bonds and does not have any effector functions.

References

1. Rijkers T, Peetz A, Ruther U (1994) Insertional mutagenesis in transgenic mice. *Transgenic Res* **3**: 203–215.
2. Bowes C, Li T, Frankel WN, *et al.* (1993) Localization of a retroviral element within the rd gene coding for the b-subunit of cGMP phosphodiesterase. *Proc Natl Acad Sci USA* **90**: 2855–2859.
3. Ochman H, Gerber SA, Hartl DL (1988) Genetic applications of an inverse polymerase chain reaction. *Genetics* **120**: 621–625.
4. Perucho M, Hanahan D, Lipsich L, Wigler M (1980) Isolation of the chicken thymidine kinase gene by plasmid rescue. *Nature* **285**: 207–210.
5. O'Kane CK, Gehring WJ (1987) Detection in situ of genetic regulatory elements in *Drosophila*. *Proc Natl Acad Sci USA* **84**: 9123–9127.
6. Allen ND, Cran DG, Barton SC, Hettle S, Reik W, Surani MA (1988) Transgenes as probes for active chromosomal domains in mouse development. *Nature* **333**: 852–855.
7. Gossler A, Joyner AL, Rossant J, Skarnes WC (1989) Mouse embryonic stem cells and reporter constructs to detect developmentally regulated genes. *Science* **244**: 463–465.
8. Von Melchner H, DeGregori JV, Rayburn H, Reddy S, Friedel C, Ruley HE (1992) Selective disruption of genes expressed in totipotent embryonal stem cells. *Genes & Dev* **6**: 919–927.
9. Friedrich G, Soriano P (1991) Promoter traps in embryonic stem cells: a genetic screen to identify and mutate developmental genes in mice. *Genes & Dev* **5**: 1513–1523.
10. Skarnes WC, Auerbach BA, Joyner AL (1992) A gene trap approach in mouse embryonic stem cells: the lacZ reporter is activated by splicing, reflects endogenous gene expression, and is mutagenic in mice. *Genes & Dev* **6**: 903–918.
11. Thorey IS, Muth K, Russ AP, Otte J, Reffelmann A, von Melchner H (1998) Selective disruption of genes transiently induced in differentiating mouse embryonic stem cells by using gene trap mutagenesis and site-specific recombination. *Mol Cell Biol* **18**: 3081–3088.
12. Wiles MV, Vauti F, Otte J, *et al.* (2000) Establishment of a gene-trap sequence tag library to generate mutant mice from embryonic stem cells. *Nature Genet* **24**: 13–14.
13. Zambrowicz BP, Friedrich GA, Buxton EC, Lilleberg SL, Person C, Sands AT (1998) Disruption and sequence identification of 2,000 genes in mouse embryonic stem cells. *Nature* **392**: 608–611.
14. Bronchain OJ, Hartley KO, Amaya A (1999) A gene trap approach in Xenopus. *Curr Biol* **9**: 1195–1198.
15. Rorth P (1996) A modular misexpression screen in *Drosophila* detecting tissue-specific phenotypes. *Proc Natl Acad Sci USA* **93**: 12418–12422.
16. Rorth P, Szabo K, Bailey A, *et al.* (1998) Systematic gain-of-function genetics in *Drosophila*. *Development* **125**: 1049–1057.
17. Spradling AC, Stern DM, Kiss I, Roote J, Laverty T, Rubin GM (1995) Gene disruptions using P transposable elements: an integral component of the *Drosophila* genome project. *Proc Natl Acad Sci USA* **92**: 10824–10830.
18. Spradling AC, Stern D, Beaton A, Rhem EJ, Laverty T, Mozden N, Misra S, Rubin GM (1999) The BDGP Gene Disruption Project: single P element insertions mutating 25% of vital *Drosophila* genes. *Genetics* **153**: 135–177.
19. Katsuki M, Sato M, Kimura M, Yokoyama M, Kobayashi K, Nomura T (1988) Conversion of normal behaviour to shiverer by myelin basic protein antisense cDNA in transgenic mice. *Science* **241**: 593–595.
20. Erickson RP, Lai LW, Grimes J (1993) Creating a conditional mutation of Wnt-1 by antisense transgenesis provides evidence that Wnt-1 is not essential for spermatogenesis. *Dev Biol* **14**: 274–281.

21. Munir M, Rossiter B, Caskey C (1990) Antisense RNA production in transgenic mice. *Somat Cell Mol Genet* **16**: 383–394.
22. Rossi JJ (1995) Controlled, targeted, intracellular expression of ribozymes: progress and problems. *Trends Biotechnol* **13**: 301–306.
23. James HA, Gibson I (1998) The therapeutic potential of ribozymes. *Blood* **91**: 371–382.
24. Heinrich JC, Tabler M, Louis C (1983) Attenuation of white gene expression in transgenic *Drosophila melanogaster*. Possible role of catalytic antisense RNA. *Dev Genet* **14**: 258–265.
25. Welch PJ, Barber JR, Wong-Staal F (1998) Expression of ribozymes in gene transfer systems to modulate target RNA levels. *Curr Opin Biotechnol* **9**: 486–496.
26. Muotri AR, da Veiga Pereira L, dos Reis Vasques L, Menck CFM (1999) Ribozymes and the anti-gene therapy: how a catalytic RNA can be used to inhibit gene function. *Gene* **237**: 303–310.
27. Efrat S, Lieser M, Wu Y, Fusco-DeMane D, Emran O, Surana M, Jetton T, Magnuson M, Weir G, Fleischer N (1994) Ribozyme-mediated attenuation of pancreatic β-cell glucokinase expression in transgenic mice results in impaired glucose-induced insulin secretion. *Proc Natl Acad Sci USA* **91**: 2051–2055.
28. Pal-Bhadra M, Bhadra U, Birchler JA (1997) Cosuppression in *Drosophila*: gene silencing of alcohol dehydrogenase by white-Adh transgenes is Polycomb-dependent. *Cell* **90**: 479–490.
29. Bahramian MB, Zabl H (1999) Transcriptional and posttranscriptional silencing of rodent alpha1 (I) collagen by a homologous transcriptionally self-silenced transgene. *Mol Cell Biol* **19**: 274–283.
30. Dernberg AF, Zalevsky J, Colaiacovo MP, Villeneuve AM (2000) Transgene-mediated cosuppression in the *C. elegans* germ line. *Genes & Dev* **14**: 1578–1583.
31. Fire A, Xu S, Montgomery MK, Kostas SA, Driver SE, Mello CC (1998) Potent and specific genetic interference by double stranded RNA in *Caenorhabditis elegans*. *Nature* **391**: 806–811.
32. Voinnet O (2002) RNA silencing: small RNAs as ubiquitous regulators of gene expression. *Current Opin Plant Biol* **5**: 444–451.
33. Tuschll T, Borkhardt A (2002) Small interfering RNAs: A revolutionary tool for the analysis of gene function and gene therapy. *Mol Int* **2**: 158–167.
34. Kennerdell JR, Carthew RW (2000) Heritable gene silencing in *Drosophila* using double stranded RNA. *Nature Biotechnol* **17**: 896–898.
35. Wianny F, Zernicka-Goetz M (2000) Specific interference with gene function by double stranded RNA in early mouse development. *Nature Cell Biol* **2**: 70–75.
36. Tabara H, Grishok A, Mello CC (1998) RNAi in *C. elegans*: soaking in the genome sequence. *Science* **282**: 430–431.
37. Timmons L, Fire A (1998) Specific interference by ingested dsRNA. *Nature* **395**: 854.
38. Hammond SM, Caudy AA, Hannon GJ (2001) Post-transcriptional gene silencing by double-stranded RNA. *Nature Rev Genet* **2**: 110–119.
39. Gonczy P, Echerverri G, Oegema K, *et al.* (2000) Functional genomic analysis of cell division in *C. elegans* using RNAi of genes on chromosome III. *Nature* **408**: 331–336.
40. Maeda I, Kohara Y, Yamamoto M, Sugimoto A (2001) Large-scale analysis of gene function in *Caenorhabditis elegans* by high-throughput RNAi. *Curr Biol* **11**: 171–176.
41. Fraser AG, Kamath RS, Zipperlen P, Martinez-Campos M, Sohrmann M, Ahringer J (2000) Functional genomic analysis of *C. elegans* chromosome I by systematic RNA interference. *Nature* **408**: 325–330.
42. Kamath RS, Fraser AG, Dong Y *et al.* (2003) Systematic functional analysis of the *Caenorhabditis elegans* genome using RNAi. *Nature* **421**: 231–237.

43. Morgan DO, Roth RA (1988) Analysis of intracellular protein function by antibody injection. *Immunol Today* **9**: 84–88.
44. Richardson JH, Marasco WA (1995) Intracellular antibodies: development and therapeutic potential. *Trends Biotechnol* **13**: 306–310.
45. Richardson JH, Hofmann W, Sodroski JG, Marasco WA (1998) Intrabody-mediated knockout of the high affinity IL-2 recdeptor in primary human T cells using a bicistronic lentivirus vector. *Gene Therapy* **5**: 635–644.
46. Beerli RR, Wels W, Hynes NE (1994) Intracellular expression of single chain antibodies reverts ErbB-2 transformation. *J Biol Chem* **269**: 23931–23936.
47. Rondon IJ, Marasco WA (1997) Intracellular antibodies (intrabodies) for gene therapy of infectious diseases. *Annu Rev Microbiol* **51**: 257–283.
48. Hemmati-Brivanlou A, Melton DA (1992) A truncated activin receptor inhibits mesoderm induction and formation of axial structures in *Xenopus* embryos. *Nature* **359**: 609–614.

Protocols

CONTENTS

MATERIALS

1M Tris.HCl (pH 7.5)

In vitro transcription vector (e.g. pBluescript series) containing transgene of interest

Restriction enzymes to linearize transcription vector at each end of the transgene

Proteinase K solution

20% sodium dodecylsulfate in DEPC-treated water

DEPC-treated water

Phenol/chloroform (1:1)

Chloroform

3M sodium acetate

Ethanol (70%, absolute)

MilliQ water

ATP, UTP, CTP and GTP, 100 mM solutions

RNA polymerase compatible with transcription vector, and appropriate buffer

RNAse-free DNAse I

5M ammonium acetate in 0.1M EDTA

Isopropyl alcohol

Annealing buffer (20 mM Tris, pH 7.4; 0.2 mM EDTA)

RNase A (1 mg ml^{-1})

RNase T1 (2 mg ml^{-1})

Protocol 9.1: *In vitro* synthesis of dsRNA

1. Prepare the sequence for expression in a suitable *in vitro* transcription vector such as pBluescriptII, which has dual opposing promoters. Separate the plasmid into two aliquots and linearize at opposite ends of the transgene as appropriate, then carry out the following process in tandem with the appropriate restriction enzymes and RNA polymerases.

2. Linearize 10 µg of the plasmid by digestion for 2 h with 10 units of the appropriate restriction enzyme.

3. Transfer the DNA to a fresh tube and add 2 µl proteinase K (25 mg ml⁻¹), 10 µl 20% SDS, 360 µl water (DEPC-treated). Vortex briefly and incubate for 20 min at 37°C.

4. Extract once with phenol/chloroform to remove proteinase K, and once with chloroform to remove phenol.

5. Transfer aqueous phase to a fresh tube, and add 40 µl 3M sodium acetate and 2.5 volumes of absolute ethanol.

6. Recover the DNA by centrifugation at 13 000 *g* for 15 min. Remove supernatant and wash pellet with 0.5 ml of 70% ethanol at room temperature. Centrifuge for 1 min as above, remove supernatant and allow pellet to dry.

7. Dissolve the pellet in 20 µl MilliQ water.

8. Mix 1 µg (2 µl) of linearized DNA with the components of the transcription reaction (ATP, UTP, CTP, GTP, 10x reaction buffer, 10× enzyme mix) in 20 µl reaction volume made up in MilliQ water.

9. Allow transcription to proceed at 37°C for 2 h.

10. Remove the DNA template by adding 2 IU RNAse-free DNAse I and incubate for 20 min at 37°C.

11. Add 115 µl water and 15 µl 5 M ammonium acetate, 0.1M EDTA to stop the reaction. Extract with phenol/chloroform as above.

12. Purify the single-stranded RNA by adding 150 µl (1 volume) isopropyl alcohol. Precipitate by freezing at –80°C for 1 h or –20°C overnight. Resuspend the pellet in 10 µl water.

13. Determine the concentration of each RNA preparation and adjust to 4 µg µl⁻¹.

14. Prepare the annealing mixture by combining 4 µl sense RNA, 4 µl antisense RNA and 8 µl 2× annealing buffer (20 mM Tris

pH 7.4, 0.2 mM EDTA). Denature at 70°C for 10 min and incubate at 37°C for 1 h.

15. Remove the single-stranded RNAs by treatment with RNase digestion mix. To 16 μl of the annealed RNA mix, add 35 μl 1 M Tris (pH 7.5) 3.5 μl 0.5 M EDTA, 21 μl 5 M NaCl, 0.35 μl RNAseA (1 mg ml^{-1} stock), 0.35 μl RNAse T1 (2 mg ml^{-1} stock) and bring up to 350 μl with water. Vortex and incubate for 30 min at 37°C.

16. Treat with proteinase K and extract with phenol/chloroform and chloroform as in steps 3 and 4.

17. Precipitate dsRNA by adding 40 μl 3 M sodium acetate and 1 ml ethanol and incubating at –80°C for 30 min. Centrifuge at 13 000 *g*, wash pellet with 80% ethanol and dry.

18. Resuspend pellet in 10 μl MilliQ water.

19. The RNA is prepared as above is suitable for injection into C. *elegans* or *Drosophila*, mouse oocytes and embryos, and for transfection into ES cells. It is not suitable for the majority of mammalian cell lines which require pre-prepared siRNA (*Protocol 9.2*).

Protocol 9.2: Transfection of mammalian cells with siRNA

MATERIALS

Mammalian cell line of interest (make sure that cells are healthy and greater than 90% viable before transfection)

siRNA of interest (20 pmol μl^{-1})

Lipofectamine 2000 (store at +4°C until use)

Opti-MEM I Reduced Serum Medium (Invitrogen, Catalog no. 31985-062; prewarmed)

24-well tissue culture plates and other tissue culture supplies.

METHOD

1. One day before transfection, plate cells in 0.5 ml of growth medium without antibiotics so that they will be 30–50% confluent at the time of transfection.

2. For each transfection sample, dilute the appropriate amount of siRNA in 50 µl Opti-MEM I Reduced Serum Medium without serum (or other medium without serum). Mix gently.

3. Mix Lipofectamine 2000 gently before use, then dilute the appropriate amount in 50 µl of Opti-MEM I Medium (or other medium without serum). Mix gently and incubate for 5 minutes at room temperature. Note: Combine the diluted Lipofectamine 2000 with the diluted siRNA within 30 minutes as longer incubation times may decrease activity. If DMEM is used as a diluent for the Lipofectamine 2000, mix with the diluted siRNA within 5 minutes.

4. After the 5 minute incubation, combine the diluted siRNA with the diluted Lipofectamine 2000 (total volume is 100 µl). Mix gently and incubate for 20 minutes at room temperature to allow the siRNA:Lipofectamine 2000 complexes to form.

5. Add 100 µl of siRNA:Lipofectamine 2000 complexes to each well. Mix gently by rocking the plate back and forth.

6. Incubate the cells at 37°C in a CO_2 incubator for 24–72 hours until they are ready to assay for gene knockdown. It is generally not necessary to remove the complexes or change the medium; however, growth medium may be replaced after 4–6 hours without loss of transfection activity.

Applications of gene transfer in animals

10

10.1 Introduction

The majority of this book has focused on principles and procedures for the genetic modification of animal cells and animals, but in this brief final chapter we will survey the major applications of this technology, and some of the breakthroughs that have been achieved. The applications of gene transfer to animal (including human) cells can be divided into seven major categories:

- The study of individual gene functions and aspects of gene regulation in an authentic genetic background.
- The high-throughput analysis of gene functions (functional genomics).
- The production of recombinant therapeutic proteins.
- The creation of disease models and models for drug testing.
- The improvement of livestock traits.
- The production of tissues and organs compatible with medical applications.
- Gene medicine – the treatment or prevention of disease.

10.2 Gene transfer for the functional analysis of individual genes

The most direct way to address the function of a gene in the context of a cell or whole organism is to disable the gene or its product, and study the effects. In mice and flies, gene targeting is a suitable approach, while in the worm and in many cell lines, RNAi would now be the method of choice. For those organisms whose genomes have been saturated with mutations, insertional or otherwise, it may be possible to isolate a mutation affecting the gene of interest. In many cases, such insertional mutations, knockouts and RNAi phenocopies are extremely informative and can reveal a great deal about the function of the gene.

In a surprisingly large number of cases, however, loss-of-function mutations appear to have very little phenotypic effect because of genetic redundancy, a phenomenon in which the functions of one gene are replaced by those of another if the first gene is inactivated. In cell lines, such redundancy affects many components of the cell cycle, so loss of expression has little effect on the cellular phenotype. However, where loss of function effects are uninformative, gain-of-function effects can provide useful information. Many oncogenes have been discovered not because their loss of function results in a failure of proliferation, but because a gain for function leads to excessive proliferation. A similar example from developmental biology is the series of mouse knockouts of muscle-specific transcriptional regulators, especially the early myogenic bHLH family, which includes Myogenin, MyoD, Myf-5 and MRF4. Before the knockouts were carried out, the important role of these regulators had been established in transfection experiments. The introduction of cDNAs encoding any of these regulators into fibroblasts and various other cell lines caused them to start to express muscle-specific markers and, in many cases, to undergo myogenic changes in morphology. Investigators therefore got a surprise when homozygous knockout mice lacking the MyoD gene product were entirely normal, and Myf-5 knockout mice showed only the subtlest of phenotypes.

Further work began to unravel the regulatory system controlling these myogenic factors and offered an explanation for the perplexing results of knockout experiments. It was shown that MyoD and Myf-5 were part of an autoregulatory loop, where each transcription factor repressed the gene encoding the other. Therefore, loss of MyoD released the *myf-5* gene from repression and allowed more Myf-5 to be produced, while loss of Myf-5 similarly released the *myoD1* gene from repression. The investigators predicted that simultaneous knockouts of both genes would show a much more dramatic phenotype and this turned out to be just the case: mice lacking both gene products failed to initiate muscle development and died just after birth from asphyxiation.

Whereas in cells it is usually only possible to generate gain-of-function effects by overexpression, more sophisticated approaches are available in transgenic animals. A transgene corresponding to a given endogenous gene can be placed under a heterologous promoter so that it is expressed ectopically, i.e. outside its normal spatial or temporal domains. This type of *mis-expression* experiment has been carried out for many of the genes governing early embryonic development, resulting in specific patterning abnormalities that provided useful clues as to the normal developmental functions of those genes. It has even proven possible to use gene targeting in mice to replace the coding region of one gene with that of another, and thus 'borrow' the regulatory elements of the first gene to drive expression of the second. Hanks and colleagues used this knock-in strategy to express the mouse *engrailed-2* gene under the control elements for the closely related *engrailed-1* gene (*Figure 10.1*). As is the case for the myogenic regulators discussed above, the engrailed genes encode important developmental transcription factors but *engrailed-2* knockout mice were more-or-less normal, suggesting that another gene, probably *engrailed-1*, provided compensatory functions when the *engrailed-2* gene was inactive. By replacing *engrailed-1* with *engrailed-2*, Hanks and colleagues were able to show that the resulting mice were phenotypically normal, suggesting the two genes were indeed functionally equivalent[1]. (See references (1,2).)

---

1 Mouse *engrailed-1* knockouts showed severe developmental brain defects, probably reflecting the fact that *engrailed-1* is switched on several hours before *engrailed-2* and in this brief period there is no redundancy to rely on. Similarly, the expression of *myf-5* begins just before that of *myoD1* in mice, and this probably explains the mild developmental phenotype of *myf-5* knockout.

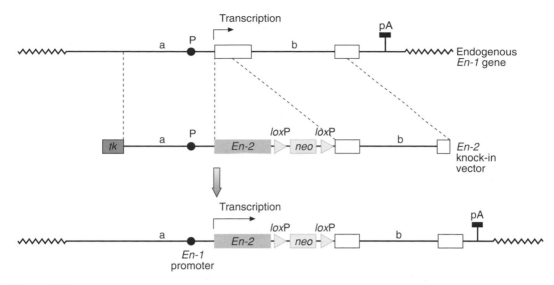

Figure 10.1

The knock-in method replaces the coding sequence of one gene with that of another. In the experiment shown, the *engrailed-1* gene (*En-1*) is replaced with the equivalent segment of *engrailed-2* (*En-2*). The targeting vector contains a homology region comprising an upstream sequence A, which contains the *En-1* promoter, and an internal segment B which spans the 3′ end of exon 1, the single intron and the 5′ coding sequence of exon 2. Separating these two sequences is the coding sequence of the *En-2* gene. A marker cassette is also introduced for selection of targeting events in ES cells. The knocked-in *En-2* gene comes under the control of the *En-1* promoter. Taken from Strachan *et al*, *Human Molecular Genetics*, Copyright (2004), Garland Science.

10.3 Gene transfer for the study of gene regulation

One of the earliest uses for transformed cell lines and transgenic organisms was the analysis of gene regulation, i.e. the identification of sequences required for specific attributes of gene expression. As discussed in Chapter 1, this is usually achieved by joining the regulatory elements of the gene under investigation to a suitable reporter gene, and assaying for the activity of the reporter protein in the presence of different amounts of regulatory sequence. A general method for mapping regulatory elements in cell lines and transgenics is to generate stepwise deletions and test the constructs in a systematic manner, as shown in the example in *Figure 10.2*. The use of different cell lines, some of which express the corresponding endogenous gene and some of which do not, may provide evidence for promoter elements that permit gene expression in permissive cell lines and prevent gene expression in non-permissive cell lines. Similarly, elements for inducible gene expression can be located by assaying different promoter constructs in the same cell line either in the presence or absence of induction. Studying such reporter expression patterns in transgenic animals can be more informative than the use of cell lines, as it becomes possible not only to identify elements responsible for quantitative changes in gene expression, but also those responsible for changes in spatial or temporal gene expression profiles. However, data must be collected from a number

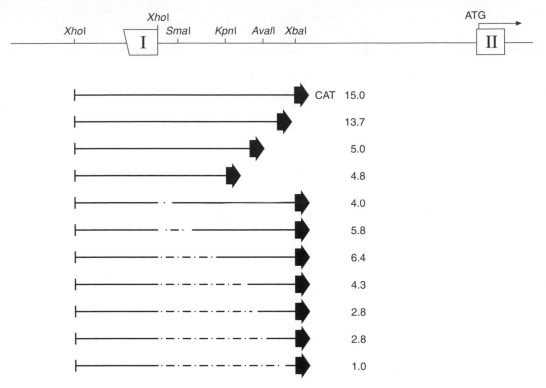

Figure 10.2

Strategy for the analysis of gene regulation. The top panel shows a restriction map of the rat-neuron-specific enolase (*NSE*) gene, showing the first and second exons as boxes. The first exon has a variable transcriptional start site, represented by the sloping edge. The translational start site is within the second exon. A set of stepwise deletion constructs was generated, in which 300bp of the *NSE* promoter (up to the *Xho*I site), the first exon and variable amounts of the first intron were used to drive the *cat* reporter gene. The numbers represent the level of CAT activity compared to the basal construct, which has 300bp of the promoter and the first exon, but no intron material at all, and has a relative CAT activity of 1.0. The construct containing the first half of the intron (up to the *Xba*I site) shows 15× more activity than the basal construct indicating the presence of important cis-acting regulatory elements in the intron. Reprinted from Molecular Brain Research, Vol 28, No. 1, Sakimura *et al*, "*Upstream and intron . . .*", Pages 19–28, Copyright (1995), with permission from Elsevier.

of independently derived transgenic lines since aberrant expression patterns could be produced by position effects and other epigenetic phenomena as discussed in Chapter 8. (See reference (3).)

10.4 Applications of gene transfer in functional genomics

Functional genomics is the large-scale analysis of gene function, a scientific discipline which has necessarily evolved from the genome-sequencing era due to the massive numbers of functionally uncharacterized genes arising from various sequencing projects. From the genomes that have been sequenced thus far, it appears that, regardless of how well characterized the organism is, about 30% of genes have no assigned function and another

30% have only loosely assigned functions based on sequence or structural similarities to known proteins – assignments such as 'kinase' or 'DNA-binding protein' which do not reveal either precise cellular functions or roles in the whole body. Gene transfer has become an indispensable tool for functional genomics because it provides several high-throughput methods for the analysis of gene function. Examples include large-scale systematic mutagenesis (either by homologous recombination as has been carried out in the yeast) or by insertional mutagenesis, gene trapping, activation trapping or genome-wide RNA interference. The principles behind these various methods were discussed in Chapter 9.

Although there has been no systematic approach to gene targeting in the mouse, so many gene knockout experiments have now been carried out that Internet databases have been established to catalog all the results, categorizing them by gene, targeting method, phenotype and other criteria. Similar databases are available cataloging *Drosophila* gene and activation trap libraries and the results from *C. elegans* RNAi projects. Some resources are listed in *Table 10.1*. The breathtaking scale of the RNAi experiments in *C. elegans* demonstrates the power of gene transfer approaches in functional genomics. Several large-scale pilot experiments have been carried out in *C. elegans*, involving the synthesis of thousands of dsRNA molecules and their systematic administration to worms either by microinjection, soaking or feeding. This has culminated in a recent, genome-wide screen in which nearly 17 000 bacterial strains were generated, each expressing a different dsRNA, representing 86% of the genes in the *C. elegans* genome. (See references (4–9).)

Table 10.1 Internet resources for mammalian transgenesis and mutagenesis, and for large-scale animal gene trap and RNAi screens

Resource	URL
Transgenic mouse and targeted mutagenesis resources	
TBASE, database of transgenic and targeted mice, maintained by the Jackson Laboratory	http://tbase.jax.org
Frontiers in Science gene knockout database	http://bioscience.org/knockout/knochome.htm
BioMedNet mouse knockout database	http://biomednet.com/db/mkmd
Gene trap resources (for mouse and Drosophila)	
German Gene Trap Consortium	http://tikus.gsf.de
Lexicon Genetics	http://www.lexgen.com/omnibank/omnibank.htm
Berkeley *Drosophila* Genome Project	http://www.fruitfly.org/p_disrupt/index.html
C. elegans RNA interference resources	
General information	
RNAi	http://www.wormbase.org http://www.rnai.org

10.5 The production of recombinant therapeutic proteins

Animal cells have been used for many years to produce recombinant proteins on a commercial scale. Some of the products include antibodies (produced in hybridoma cells) and recombinant enzymes, cytokines and blood products produced predominantly in CHO (Chinese hamster ovary) and NS0 myeloma cell lines, which can be maintained in large fermentor vessels and for which validated protocols for transfection, culture, product recovery and purification can be applied under good manufacturing practice (GMP) conditions. These cells have the advantage of high yields and the production of proteins with authentic glycan structures. Despite the long and successful history of cell-based production, this system is expensive (in terms of both equipment and media requirements) and lacking in capacity. Attention has turned to the possible use of transgenic animals for the production of therapeutic proteins in milk and other body fluids because of the high yields and non-destructive harvesting of the recombinant product.

Although a transgenic mouse expressing rat growth hormone in its serum was produced by Brinster and colleagues as early as 1982, the first genuine animal 'bioreactor' was a transgenic mouse generated by Gordon and colleagues producing human tissue plasminogen activator in its milk. Proteins can be produced at high concentrations in mouse milk (e.g. 50 ng ml^{-1} for tissue plasminogen activator), but only a small volume of milk is produced by each animal. Therefore, conventional dairy animals such as sheep and goats have been investigated as alternative bioreactors, since they have larger production volumes and are more acceptable to use for pharmaceutical production because they are already used as food and are therefore generally regarded as safe. Many proteins have been produced in animal milk, a selection of which is listed in *Table 10.2*. Recombinant human antithrombin III, which is used to prevent blood clots forming in patients who have undergone heart bypass operations, was the first protein expressed in transgenic animal milk to reach commercial production, and is currently marketed by Genzyme Transgenics Corporation. The production of foreign proteins in secreted body fluids has the obvious advantage that transgenic animals can be used as a renewable source of the desirable molecule. In addition to milk, a number of other production systems have been investigated, including serum, semen, urine and hens' eggs. There has also been some success with silkworm cocoons as a production system, following transduction of expression constructs into silkworm larvae using baculovirus vectors and the microinjection of DNA into silkworm eggs. (See references (10–12).)

10.6 Disease models

Human diseases are often characterized by their effect on the whole body, but many diseases have a cellular phenotype which means that cell lines rather than animals can be used to create disease models and test drugs. For example, xeroderma pigmentosum is a disease which primarily affects the eyes and skin, but the defect at the cellular level involves a deficiency in DNA repair which can be studied in cultured cells. Diseases character-

Table 10.2 Recombinant proteins produced in the milk of transgenic animals

Coding sequence		Transgenic species	Promotor region	
Gene	Source		Gene	Source
α_1-antitrypsin	Mouse	Mice	WAP	Rabbit
α_1-antitrypsin	Human	Mice	β-lactoglobulin	Ovine
α_1-antitrypsin	Human	Sheep	β-lactoglobulin	Ovine
α-glucosidase	Human	Mice	α_{s1}-casein	Bvine
α-lactalbumin	Bovine	Mice	α-lactalbumin	Bovine
α-lactalbumin	Goat	Mice	α-lactalbumin	Goat
α-lactalbumin	Guinea-pig	Mice	α-lactalbumin	Guinea-pig
α-lactalbumin	Human	Rats	α-lactalbumin	Human
α_{s1}-casein	Bovine	Mice	α_{s1}-casein	Bovine
Anti-CD6 antibodies	Mouse/human	Mice	WAP	Rabbit
Antithrombin III	Human	Goat	J-casein	Goat
β-casein	Bovine	Mice	β-casein	Bovine
β-casein	Bovine	Mice	α-lactalbumin	Bovine
β-casein	Goat	Mice	β-casein	Goat
β-casein	Rat	Mice	β-casein	Rat
β-interferon	Human	Mice	WAP	Mouse
β-lactoglobulin	Ovine	Mice	β-lactoglobulin	Ovine
γ-interferon	Human	Mice	β-lactoglobulin	Ovine
κ-casein	Bovine	Mice	β-casein	Goat
κ-casein	Goat	Mice	β-casein	Goat
CFTR	Human	Mice	β-casein	Goat
EPO	Human	Mice	β-lactoglobulin	Bovine
EPO	Human	Rabbits	β-lactoglobulin	Bovine
Factor VIII	Human	Sheep	β-lactoglobulin	Ovine
Factor IX	Human	Mice	β-lactoglobulin	Ovine
Factor IX	Human	Sheep	β-lactoglobulin	Ovine
Fibrinogen	Human	Mice	WAP	Mouse
FSH	Bovine	Mice	J-casein	Rat
GM-CSF	Human	Mice	α_{s1}-casein	Bovine
Growth hormone	Bovine	Mice	WAP	Rat
Growth hormone	Human	Mice	J-casein	Rat
Hepatitis B surface antigen	Human	Goat	α_{s1}-casein	Bovine
IGF-1	Human	Rabbits	α_{s1}-casein	Bovine
Interleukin-2	Human	Rabbits	β-casein	Rabbit
Lactoferrin	Human	Mice	α_{s1}-casein	Bovine
Lactoferrin	Human	Cattle	α_{s1}-casein	Bovine
Lysozyme	Human	Mice	α_{s1}-casein	Bovine
Protein C	Human	Mice	WAP	Mouse
Protein C	Human	Pigs	WAP	Mouse
Serum albumin	Human	Mice	β-lactoglobulin	Ovine
Superoxide dismutase	Human	Mice	β-lactoglobulin	Ovine
Superoxide dismutase	Human	Mice	WAP	Mouse
Surfactant protein B	Human	Mice	WAP	Rat
TAP	Human	Mice	WAP	Rat
t-PA	Human	Mice	WAP	Mouse
t-PA	Human	Mice	α_{s1}-casein	Bovine
t-PA	Human	Rabbits	α_{s1}-casein	Bovine
t-PA	Human	Goats	WAP	Mouse
Trophoblastin	Sheep	Mice	α-lactalbumin	Bovine
Urokinase	Human	Mice	α_{s1}-casein	Bovine
WAP	Mouse	Mice	WAP	Mouse
WAP	Rat	Mice	WAP	Rat
WAP	Mouse	Pigs	WAP	Mouse
WAP	Mouse	Sheep	WAP	Mouse

Reprinted from Livestock Production Science, Vol. 59, Wall RJ *et al*, "*Biotechnology* ...", Copyright (1999), with permission from Elsevier.

ized by a gain of function can be modeled by the transient or stable expression of a dominant transgene, and this approach has been used very widely to study oncogenes and test drugs to suppress their activities. With the advent of RNA interference-based methods for gene inhibition in mammalian cells, it is now becoming possible to model loss-of-function diseases in a similarly straightforward manner, simply by transfection with the appropriate siRNA.

While cell models can be used as a first line of investigation for some diseases, animal models are required to study diseases in a whole-organism context. Some animal disease models arise spontaneously, or can be induced by random chemical mutation or irradiation. However, gene transfer provides a more direct route because specific genetic lesions can be introduced into the genome. The mouse is one of the best organisms for disease modeling because its genome is perhaps the easiest of all animals to manipulate, yet its overall genetic and physiological properties are similar to humans.

The majority of human genetic diseases result from the loss of a gene function and these are usually modeled by creating an equivalent knockout mouse by gene targeting. The first step in this procedure is to identify the mouse gene which corresponds to the human gene involved in the disease. Such a functionally equivalent gene is described as an ortholog. In many cases, it is sufficient to use the standard gene targeting methodology and disrupt the gene by inserting a large cassette into an essential exon. This creates a null allele, i.e. expression is completely abolished. However, in other cases it is necessary to model the precise mutation found in the human disease, as this may be a subtle mutation, such as a point mutation that alters the structure of the resulting protein but maintains the normal expression level. Mouse models of human diseases produced by gene targeting do not always mirror the human disease. In some cases, this is because the targeting has not achieved the desired result (e.g. splicing and sometimes reconstitute the wild-type gene product and restore some functionality to the targeted locus) while in other cases, it may reflect genuine differences between murine and human physiology.

Diseases which are caused by a gain of gene function can be modeled in mice (and indeed other mammals) by introducing a transgene by pronuclear microinjection. This is because dominantly acting gene products will exert their effects even in the presence of the wild-type gene product, and therefore inactivation of the endogenous gene is generally not required. To model such disorders, a dominant mutant version of the affected gene must be cloned and introduced into the mouse (the human gene is often sufficient) or an equivalent mutation can be introduced into the cloned mouse ortholog by *in vitro* mutagenesis. Examples of diseases that have been modeled successfully using this strategy include cancers and expanded triplet repeat disorders such as Huntington's disease and spinocerebellar ataxia type I. (See references (13–22).)

10.7 Improvement of livestock traits

Most of the desirable traits of farm animals are quantitative traits, which have been improved over thousands of years through selective breeding.

The application of gene transfer to the improvement of livestock traits is therefore in its infancy and there are few examples, as yet, where single transgenes have had a significant impact. One exception already described in this book is the disruption of the *PRP* gene in sheep by gene targeting in somatic cells, followed by nuclear transfer to an enucleated egg. By inactivating the *PRP* gene, the sheep are expected to be resistant to the prion disease scrapie. Similar experiments have been conducted in cows in an attempt to make them resistant to bovine spongiform encephalopathy.

One area in which gene transfer has been widely used to improve animal traits is the genetic modification of fish. Some investigators have focused on disease resistance, but most progress has been made with two groups of traits: growth rate and resistance to freezing. Currently, at least one US biotechnology company is close to commercializing transgenic fish engineered to grow more rapidly: AquaBounty Farms (Waltham, MA) is waiting for approval from the US Food and Drug Administration for Atlantic salmon carrying a growth hormone gene from Chinook salmon, and this is likely to be the first genetically modified animal approved for human consumption. A large number of growth-enhanced fish have been reported, and in the best case the growth rate was increased 35-fold. Many Arctic fish express protective proteins in the circulation which inhibit freezing in cold waters, and the transfer of such genes into commercial foodstocks such as salmon could allow the northward expansion of fish farms. (See references (23,34).)

10.8 The production of human-compatible tissues for transplantation

Transplantation is the best treatment available for organ failure but there is always a shortage of organ donors, making it difficult to find organs that are compatible with the patient. Transgenic animals, particularly pigs, could be engineered to produce organs customized for human recipients, reducing the time patients have to wait for transplants and therefore saving many lives. The process of transferring organs between species is known as xenotransplantation.

Although the ethical issues surrounding xenotransplantation are hotly debated, there are also several technical hurdles that need to be overcome before the technique becomes a reality. The biggest problem is *hyperacute rejection*, which is caused by the host immune system recognizing foreign antigens on the donor organ and initiating an immune response against it, a response that includes the production of antibodies and activation of the complement system. In both cases, the major trigger is a disaccharide group (Gal-α(1,3)-Gal) known as the gal epitope, which is present on the surface of all cells of all mammals with the exception of Old World monkeys, apes and humans. Several transgenic strategies have been devised to address this problem either by preventing the gal epitope from being synthesized, expressing enzymes to convert the gal epitope into one found naturally in humans or equipping the porcine cells with complement-inactivating protein to block the immune response in the recipient. In 2002, two research groups independently reported the production of gene knockout pigs in which the gene encoding α(1,3) galactosyltransferase, the enzyme responsible for synthesizing the gal epitope, had been disrupted by homologous recombination[2]. Although only one copy of the gene was disrupted and the cells in these animals still bore the gal epitope, these reports confirmed the technical possibility of generating full homozygous knockouts. In 2003, pigs completely lacking the α-1,3-galactosyltransferase gene were reported, although the second allele was shown to have undergone a spontaneous mutation rather than a second round of gene targeting. (See references (25–30).)

10.9 Gene medicine

Gene medicine encompasses any approach involving gene transfer, which is used to treat or prevent disease. It can be divided into two major areas, the use of DNA vaccines and the use of gene therapy.

DNA vaccines differ from conventional vaccines in that the administered vaccine does not contain the antigen which raises the immune response. Instead, the vaccine is an expression construct which encodes the antigen. The body does not raise antibodies against the introduced DNA, but against the product it encodes. The DNA vaccine produces an intracellular antigen, which is presented to the immune system and results in the production of antibodies.

DNA vaccines have several advantages over conventional vaccines, not least of which is the fact that all DNA vaccines can be transported, stored and administered in a similar fashion. Other genes, encoding proteins which stimulate the immune system can be cotransferred with the vaccine, and if bacterial DNA elements are present in the delivery vector, this can also stimulate the immune system. DNA vaccines have been developed against a range of viruses, including measles, HIV, Ebola virus and influenza virus, as well as bacterial pathogens such as tuberculosis. A large number of such vaccines are currently undergoing clinical trials.

Gene therapy differs from DNA vaccination in that the disease is already established when the treatment begins, and the aim is to alleviate the symptoms of the disease (palliative therapy) or in the ideal case to effect a full cure. Gene therapy can be used to treat inherited genetic disorders, sporadic genetic diseases such as cancer and even infectious diseases. There are many different variations on gene therapy, which have been developed to treat different types of disease. One distinction is made according to how the cells are transformed – gene transfer can be achieved using viral vectors or non-viral delivery methods. Viral vectors are preferred because they are more efficient than non-viral methods, particularly for direct *in vivo* gene delivery, but as discussed in Chapter 4 there are safety concerns with such vectors, since they have been shown to provoke allergic responses in some patients and there is a residual concern with some vectors that, despite precautions, disease-causing wild-type viruses could be produced. A further concern with integrating viruses is the possibility that they could act in the same manner as activation traps, and increase the activity of adjacent genes. This appears to be just what happened in one gene therapy trial using

2 Apparently one litter was born on Christmas Day – the piglets were appropriately named Noel, Angel, Star, Joy and Mary.

retroviruses, where two of the patients who underwent successful treatment later developed leukemia because the vector integrated next to an oncogene and increased its activity. The risks associated with viral vectors have promoted research into other delivery methods, the most popular of which include direct injection of DNA into tissues (e.g. muscle), the injection of liposome–DNA complexes into the blood and direct transfer by particle bombardment (Chapter 2).

Another distinction is made according to the status of the target cells – the patient's cells may be removed, transformed with the appropriate vector, and returned to the body, or gene transfer may be carried out on cells in situ. The *ex vivo* approach is only possible where explanted cells are amenable to culture (e.g. bone marrow cells) and has the advantage that normal culture-based selection procedures can be used. A very important distinction must be made between somatic gene therapy and germline gene therapy. The former involves gene transfer to the somatic cells of an individual and is the only form of gene therapy currently allowed. Germline gene therapy is the transfer of DNA into the human germline, essentially the production of transgenic human beings, and is currently outlawed in most countries. Although there are some compelling reasons to allow germline gene therapy in rare instances, the ethical objections are very strong, including the possibility of unanticipated consequences of gene transfer (such as insertional mutagenesis) and the fact that the transgenic individuals produced in the procedure would have no right to decide.

The actual strategy used for gene therapy depends on whether the disease-causing gene is dominant or recessive, and whether cells carrying this gene need to be killed or preserved. The available strategies are summarized in *Figure 10.3* and discussed in more detail in the following sections:

- Gene-augmentation therapy. This is used to treat diseases caused by a loss of gene function. DNA is added to the genome to supply a missing gene product.
- Gene-inhibition therapy. This is used to treat diseases caused by a gain of gene function, as well as infectious diseases. DNA is added to the genome, but its function is to inhibit the dominantly malfunctioning disease gene or a gene in the infectious pathogen.
- Gene replacement therapy. The most sophisticated approach in which the disease gene is replaced with the normal allele by gene targeting. This may become a more established procedure when the efficiency of gene targeting in human cells is improved.
- The targeted ablation of specific cells. This is used to cure diseases where the affected cells need to be destroyed, e.g. many cancers.

Gene-augmentation therapy for recessive diseases

The first gene therapy trial, which began in 1990, provides a good example of gene augmentation therapy. The trial involved Ashanti DeSilva, a four-year-old girl suffering from severe combined immunodeficiency (SCID) caused by the absence of adenosine deaminase, an enzyme required for the

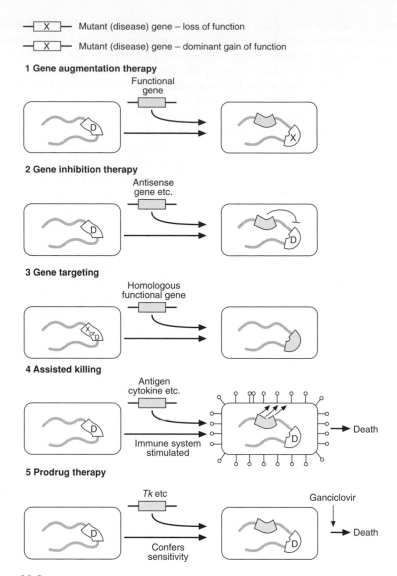

Figure 10.3

Five different gene therapy strategies for the treatment of dominant and recessive genetic diseases. From Primrose, Twyman, Old (2001) Principles of Gene Manipulation, Blackwell Science.

differentiation of T-cells. Conventional treatments for ADA-SCID include a bone marrow transplant or regular injections with the ADA enzyme, produced in genetically modified bacteria. Without these treatments ADA-SCID patients are confined to sterile containment facilities to avoid infection.

ADA deficiency was chosen as the first target for gene therapy for several key reasons:

- The disease is life threatening and conventional treatments are difficult to obtain. However, replacing the missing gene product can completely reverse the pathological symptoms of the disease.
- The basis of the diseases is the loss of function of a single gene, which has been cloned. The gene is small and easily contained in a retroviral vector.
- Levels of the enzyme differ significantly, even among normal people. This means that transgene expression does not have to be controlled precisely.
- The target cells for the therapy are lymphocytes, which are suitable for *ex vivo* modification and can be returned to the patient afterwards.

The procedure is shown in *Figure 10.4*. Bone marrow was removed from the patient and cultured under conditions to encourage the proliferation of T-cells. The cultures were then transduced with retroviral vectors carrying the *ADA* transgene, and the cells were infused back into the patient. Although successful, the effects of the treatment were transient because T-cells have a finite lifespan, and continued treatment with the ADA enzyme is therefore required. More recently, similar trials have been conducted in which bone marrow cells were used as targets for gene transfer, in the hope of transducing some of the stem cells which would repopulate the patient's bone marrow and lead to the constant production of modified T-cells. Such treatment did indeed prolong the production of ADA-positive T-cells although levels of the enzyme produced in the cells were very low. In 2002, the ADA-SCID gene therapy technique described above was combined with a method called non-myeloablative conditioning, in which bone marrow in the SCID patient is partially killed allowing the modified stem cells to proliferate. A two-year-old Palestinian child was

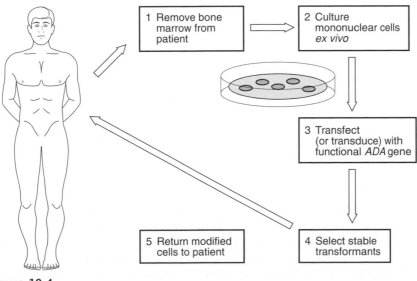

Figure 10.4

Procedure for *ex vivo* gene therapy, based on the treatment for ADA deficiency. From Primrose, Twyman, Old (2001) Principles of Gene Manipulation, Blackwell Science.

the first to undergo this modified procedure and thus far appears to be living a typical life and fighting off normal childhood infections.

X-linked SCID, a disease similar in its symptoms to ADA-SCID but this time caused by the loss of a gene on the X-chromosome, has also been treated by gene therapy. A retroviral vector carrying a functional copy of the missing interleukin receptor gene was used to transduce cultured haematopoietic stem cells, and cells transformed with the recombinant gene were transferred back into the patient. Thus far, eleven patients have been treated successfully but two of them developed leukemia following transgene integration adjacent to the *LMO2* oncogene. This ever-present risk of using retroviral vectors is exacerbated by the *ex vivo* gene therapy approach, since cells that proliferate more rapidly are more likely to be selected for transfer to the patient. The benefits and risks of retroviral gene therapy must be carefully evaluated.

Gene-therapy strategies for cancer

Cancer gene therapy provides examples of many different therapeutic strategies, several of which involve the direct transformation of tumor cells with transgenes that promote cell death or attack by the immune system. This can be achieved through the expression of cytokines, which attract killer immune cells, or the expression of foreign antigens, which are displayed on the cell surface so that the tumor cells are recognized by antibodies and destroyed. A different strategy, sometimes termed 'assisted killing' is the use of the transgenes described above to transform fibroblasts, which are easier to manipulate in culture. The fibroblasts are then injected into the patient along with tumor cells to provoke an immune response against the tumor.

Cancers caused by dominantly malfunctioning oncogenes can be treated by the direct interference with oncogene activity, which can be achieved by a number of approaches including antisense transgenes, ribozyme constructs and, more recently, short interfering RNA. Indeed RNAi may soon emerge as a novel and very potent therapeutic modality, given the highly specific way in which individual genes can be silenced. In contrast, cancers caused by loss of tumor-suppressor gene function can be addressed by introducing a functional copy of the transgene back into the cell, a similar approach to the treatment of SCID as discussed above.

Cancer cells can also be targeted by an approach known as prodrug activation therapy, in which the tumor cells are transformed with a gene encoding an enzyme that converts an innocuous prodrug into an active and highly toxic product. For example, the widely used marker gene *Tk*, which encodes the enzyme thymidine kinase, can convert the non-toxic thymidine analog ganciclovir into ganciclovir triphosphate. The latter can be incorporated into DNA in place of thymidine triphosphate, but blocks further replication leading to cell death. This approach is especially useful in cancers affecting predominantly non-dividing tissues. For example, oncoretrovirus vectors (which only infect dividing cells) can introduced the *Tk* transgene specifically into brain tumors without infecting the surrounding normal (post-mitotic) cells. Introducing ganciclovir into the brain, via the bloodstream, is harmless to the non-transformed neurons but highly toxic to the tumor cells.

References

1. Rudnicki MA, Braun B, Hinuma S, Jaenisch R (1992) Inactivation of myoD in mice leads to upregulation of the myogenic HLH gene myf-5 and results in apparently normal muscle development. *Nucleic Acids Res* **18**: 4833–4842.
2. Hanks M, Wurst W, Anson-Cartwright L, Auerbach AB, Joyner AL (1995) Rescue of the En-1 mutant phenotype by replacement of En-1with En-2. *Science* **269**: 679–682.
3. Twyman RM, Jones EA (1995) The molecular basis of neuron-specific gene expression in the mammalian nervous system. *J Neurogenet* **10**: 67–101.
4. Rorth P, Szabo K, Bailey A, *et al.* (1998) Systematic gain-of-function genetics in *Drosophila*. *Development* **125**: 1049–1057.
5. Spradling AC, Stern DM, Kiss I, Roote J, Laverty T, Rubin GM (1995) Gene disruptions using P transposable elements: an integral component of the *Drosophila* genome project. *Proc Natl Acad Sci USA* **92**: 10824–10830.
6. Gonczy P, Echerverri G, Oegema K, *et al.* (2000) Functional genomic analysis of cell division in *C. elegans* using RNAi of genes on chromosome III. *Nature* **408**: 331–336.
7. Maeda I, Kohara Y, Yamamoto M, Sugimoto A (2001) Large-scale analysis of gene function in *Caenorhabditis elegans* by high-throughput RNAi. *Curr Biol* **11**: 171–176.
8. Fraser AG, Kamath RS, Zipperlen P, Martinez-Campos M, Sohrmann M, Ahringer J (2000) Functional genomic analysis of *C. elegans* chromosome I by systematic RNA interference. *Nature* **408**: 325–330.
9. Kamath RS, Fraser AG, Dong Y, *et al.* (2003) Systematic functional analysis of the *Caenorhabditis elegans* genome using RNAi. *Nature* **421**: 231–237.
10. Wall RJ (1999) Biotechnology for the production of modified and innovative animal products: transgenic livestock bioresactors. *Livestock Prod Sci* **59**: 243–255.
11. Houdebine LM (2000) Transgenic animal bioreactors. *Transgenic Res* **9**: 305–320.
12. Dyck MK, Lacroix D, Pothier F, Sirard MA (2003) Making recombinant proteins in animals – different systems, different applications. *Trends Biotechnol* **21**: 394–399.
13. Muller U (1999) Ten years of gene targeting: targeted mousemutants, from vector design to phenotype analysis. *Mech Devel* **82**: 3–21.
14. Bedell MA, Jenkins NA, Copeland NG (1997) Mouse models of human disease. Part II. Recent progress and future directions. *Genes Dev* **11**: 11–43.
15. Clarke AR (1994) Murine genetic models of human disease. *Curr Opin Genet Dev* **4**: 453–460.
16. Erickson RP (1996) Mouse models of human genetic disease: which mouse is more like a man? *Bioessays* **18**: 993–998.
17. Ghebranious N, Donehower LA (1998) Mouse models in tumor suppression. *Oncogene* **17**: 3385–3400.
18. Macleod KF, Jacks T (1999) Insights into cancer from transgenic mouse models. *J Pathol* **187**: 43–60.
19. Smithies O (1993) Animal models of human genetic diseases. *Trends Genet* **9**: 112–116.
20. Bynshaw-Boris A (1996) Model mice and human disease. *Nature Genet* **13**: 259–260.
21. Shastry BS (1998) Gene disruption in mice: models of development and disease. *Mol Cell Biochem* **181**: 163–179.
22. Dooley K, Zon LI (2000) Zebrafish: a model system for the study of human disease. *Current Opin Genet Dev* **10**: 252–256.

23. Denning C, Burl S, Ainslie A, *et al.* (2001) Deletion of the alpha(1,3)galactosyl transferase (GGTA1) gene and the prion protein (PrP) gene in sheep. *Nature Biotechnology* **19**: 559–562.

24. Zbikowska HM (2003) Fish can be first – advances in fish transgenesis for commercial applications. *Transgenic Res* **12**(4): 379–389.

25. Lambrigts D, Sachs DH, Cooper DKC (1998) Discordant organ xenotransplantation in primates: world experience and current status. *Transplantation* **66**: 547–561.

26. Sandrin MS, McKenzie IFC (1999) Recent advances in xenotransplantation. *Curr Opin Immunol* **11**: 527–531.

27. Logan JS (2000) Prospects for xenotransplantation. *Curr Opin Immunol* **12**: 563–568.

28. Dai Y, Vaught TD, Boone J, *et al.* (2002) Targeted disruption of the alpha1,3-galactosyltransferase gene in cloned pigs. *Nature Biotechnology* **20**: 251–255.

29. Lai L, Kolber-Simonds D, Park KW, *et al.* (2002) Production of alpha-1,3-galactosyltransferase knockout pigs by nuclear transfer cloning. *Science* **295**: 1089–1092.

30. Phelps CJ, Koike C, Vaught TD, *et al.* (2003) Production of alpha 1,3-galactosyltransferase-deficient pigs. *Science* **299**: 411–414.

31. Reyes-Sandoval A, Ertl HC (2001) DNA vaccines. *Curr Mol Med* **1**: 217–243.

32. Kay MA, Glorioso JC, Naldini L (2001) Viral vectors for gene therapy: the art of turning infectious agents into vehicles of therapeutics. *Nature Med* **7**: 33–40.

33. Somia N, Verma IM (2000) Gene therapy: trials and tribulations. *Nat Rev Genetics* **1**: 91–99.

34. Daley GQ (2002) Prospects for stem cell therapeutics: myths and medicines. *Curr Opin Genet Dev* **12**: 607–613.

35. Gordon JW (1999) Genetic enhancement in humans. *Science* **283**: 2023–2024.

36. Tuschl T, Borkhardt A (2002) Small interfering RNAs: A revolutionary tool for the analysis of gene function and gene therapy. *Mol Inter* **2**: 158–167.

Index